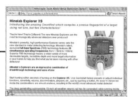

CONTENTS

Editor
Greg Payne

Design Editor
Liz Wright

Origination
Sally Robinson

Published by
Greenlight Publishing
The Publishing House, 119 Newland Street
Witham, Essex CM8 1WF
Tel: 01376 521900 **Fax:** 01376 521901
mail@greenlightpublishing.co.uk
www.greenlightpublishing.co.uk

Printed in Great Britain

ISBN 1 897738 188
© 2004 Greenlight Publishing

AUTHOR PROFILES

JULIAN EVAN-HART

I was born in 1962 at Welwyn Garden City. My family then moved to the small village of Weston, also in Hertfordshire. It was here that my passion for history, wildlife and the collecting of all things interesting started.

From a young age collecting was my pastime, and items such as pottery fragments from the local Roman villa, fossils, and other curiosities all found their way into my "museum". At the tender age of six I found a mammoth tooth lying in the recently excavated soil from some road works, to add to my ever-growing collection.

While attending St George's School in Harpenden in the mid-1970s I heard for the first time of a gadget called a metal detector. Despite being intrigued by these devices it was not until the late 1980s that I would actually be able to purchase such a machine.

I found that other people in and around Stevenage where I now live shared my passion for metal detecting. Therefore a natural progression was to get together and form a group. This achieved, we call ourselves "The Pastfinders".

To say that I am interested in metal detecting is definitely an understatement: for me it's a passion. It seems that I am always researching for Roman villas or to find the spot where a Second World War aeroplane crashed. Metal detecting has enriched my life more than any other of my past hobbies.

DAVE STUCKEY

Born in Hannover, West Germany, in 1956, I then spent most of my childhood travelling the world as part of a service family. My formative years were spent in such places as Malaya and Singapore. After finishing school in Gutersloh, again in West Germany, I then joined the Royal Navy as a Seaman Gunner in 1973. It was during my years in the RN that I became interested in the hobby of metal detecting, after seeing people with these rather curious devices pacing up and down the beaches of Southsea, near Portsmouth.

I went up to one of these guys and asked him what he was looking for. He promptly pulled out a neck chain that held dozens of gold rings, all of which he had found on that particular beach! I was so impressed that I decided to get one of these machines for myself.

I ordered my first detector, a C-Scope IB300, from Joan Allens in 1976 and haven't looked back since.

After teaming up with Julian Evan-Hart in the mid-1990s, we formed a group known as "The Pastfinders", which now has seven members in total.

Since taking up metal detecting, I have developed an enormous fascination for history and have even taken part in archaeological excavations in my own locality. I have also built up a good relationship with the Archaeological Department of Hertfordshire County Council, with whom I liase regularly.

INTRODUCTION

Did you find the subject of history tediously uninteresting when you were at school? Were you bored to tears when being made to learn about our past from chalkboards and textbooks? If you were, then you certainly weren't alone.

Perhaps, on the other hand, those visits to museums ignited some spark of interest in our past but left you feeling somewhat frustrated. You felt an overwhelming desire to touch the artefacts and coins that were once the everyday items of use by our ancestors, but those glass barriers denied you the privilege of making that physical contact with the past. Again, you certainly weren't alone.

Until about three decades ago that privilege was reserved for the lucky few such as archaeologists, museum staff, historians, and scholars.

Archaeologists, of course, would normally have been the first to touch any object that came out of the ground after having been lost or deliberately hidden for hundreds, if not thousands, of years. These finds would then have been forwarded to museums, which had the task of cleaning and conserving the artefacts prior to them being studied by experts and scholars. Only then would a select few of these treasures be put on display for the public to admire.

Towards the end of the 1960s, however, new technology appeared that would change that system and grant the privilege of handling old or ancient finds to the mainstream public. The hobby of metal detecting had been born.

Early metal detectors were quite rudimentary. Their basic design gave them the appearance of a simple transistor radio attached to a stick with a small coil on the end.

By the 1970s, however, metal detector technology had improved dramatically. The machines that began appearing on the market were vastly superior to their rather basic predecessors, and this was reflected in the number of amazing discoveries that were being made. The Water Newton hoard and the Thetford hoard are just two examples of some of the fabulous treasures that came to light in those early years.

Enthusiasm for the hobby grew and it became an increasingly popular pastime. Visitors to the coast soon became accustomed to the strange characters that could be seen pacing up and down the beach swinging their electronic "wands" in search of lost valuables.

Early "treasure hunters" who weren't fortunate enough to live near the coast were hardly disadvantaged, however. They were soon to discover that the countryside around them had an even greater potential. Britain had literally millions of acres of "virgin" farmland that had never been detected on previously. Excitement grew over the staggering number of amazing finds that were being made by the early pioneers of metal detecting. Museum cabinets soon began to fill as more and more treasures came to light.

Many began to see the hobby's potential as increasing numbers of new archaeological sites were discovered. In addition, the discovery of many previously unrecorded types of coins and artefacts improved our knowledge of Britain's past. However, there were some people who were rather less than enthusiastic about this new pastime.

Many within the archaeological community regarded metal detectorists as a threat to the nation's heritage rather than a huge potential benefit. Campaigns were launched in order to have severe restrictions imposed on the use of metal detectors; some even wanted the hobby banned altogether.

Fortunately, the rights of the general public to go exploring the countryside with metal detectors were upheld. Ancient Treasure Trove laws and Codes of Conduct were emphasised to all users of metal detectors and were accepted as a fair compromise.

Although some hardliners still exist within the archaeological community, the attitudes of many have softened in recent years. The realisation that co-operation could bring far greater benefits than confrontation has, in many cases, resulted in good working relationships in some parts of the country.

Examples of this "symbiosis" have certainly been demonstrated in recent years when major archaeological discoveries have been made by detectorists and left in situ for the archaeologists to excavate. There have even been cases where long held beliefs about the past have had to be revised owing to discoveries made by metal detectorists. A classic example of this occurred in Germany in the last decade.

In AD 9 Germanic tribesmen known as the Cherusci ambushed and massacred the armies of the Roman general Varus. Over 20,000 Roman troops and followers were slaughtered in the attack. For centuries, the site of this battle had been accepted as being somewhere in the vicinity of the modern town of Bielefeld, which is close to the Tuetoberger Vald.

Some years ago, however, a serving British Army officer using a metal detector unearthed startling evidence of a major battle some 20 miles from the town. Excavations carried out by archaeologists at the site soon revealed many fabulous relics and coins from the beginning of the 1st century AD, including a superb Roman cavalry mask. Evidence of a battle was overwhelming and soon the experts realised that the army officer had located the true site of the battle. The history books had to be re-written.

What Remains To Be Found?

Whatever your reason for taking up metal detecting, doubts may have crossed your mind as to whether there's actually still anything left out there for you to find. Well, take heart. You only have to flick through the pages of any of the major metal detecting magazines, such as **Treasure Hunting**, to see that amazing finds are still coming up in abundance. This does not mean that there is an infinite amount of material in the ground, however. Many heavily worked sites do tend to "dry up" after years of being searched and find rates diminish.

Soil conditions play a major part in finding lost coins and artefacts. The wrong conditions often cause many finds to be missed (soil conditions will be dealt with in a later chapter).

Another reason why finds are still coming up in abundance is due to the advances in metal detector design. The introduction of microprocessor technology has revolutionised metal detecting just as it has all other aspects of our lives.

In the early years, most metal detectors were simply equipped with an on/off switch and a tuning knob. By the mid-1970s some "top-of-the-range" models began sporting extra knobs to control both ground effect and discrimination (the ability to distinguish ferrous from non-ferrous metals).

Compare the models of yesteryear with what is available on the market today and you'll see a vast difference. Although there are still many fairly basic metal detectors on the market that give superb performance, many highly-sophisticated models now incorporate computer technology that gives performance and capabilities undreamt of perhaps 20 years ago. These often have a price tag to match, however.

Another reason why there is much yet to find is that there are still many hundreds of thousands of acres of land still un-searched. Most individual detectorists and detecting clubs, spend much of what time they have available searching their regular sites.

It would, of course, be almost impossible to try and estimate just how much material has already been recovered over the past 30 years and then try to estimate how much is actually left to find. But it is safe to say that more finds still remain in the soil of Britain than have so far been recovered from it.

Some of this material is buried in the ground at depths that cannot be reached without specialised and expensive equipment. Even the best machines currently on the market have limitations on their depth-seeking capabilities. You will, of course, hear many "Fishermen's tales" from other detectorists who claim that they regularly find coins at depths of 12in, for example. The truth is that the coin was originally only 3in or 4in down but in their struggle to locate it, they dug an even deeper hole and the coin kept falling down it!

We have seen remarkable improvements in the design and technology of metal detectors in recent years; some of these have increased depth-seeking capabilities to some extent. However, most detectorists are resigned to the fact that unbelievable amounts of ancient artefacts and treasures are now buried at depths that put them well beyond the reach of current technology. But who knows what the future holds?

THE DIFFERENT TYPES OF DETECTOR AVAILABLE

If you have already started to take an interest in metal detecting, then there is a good chance that you will have come across one or more of the hobby's magazines. Glancing through these, you may have been bewildered by the apparently staggering number of different machines that are currently available on the market.

Unless you have already acquired a metal detector, you will probably be wondering at this point which machine is most suitable for you. To help in making this decision you need to ask yourself the following:-

How much am I willing to spend?

How much time will I devote to using my new detector?

How good am I at adapting to new technology?

All three questions, believe it or not, are closely linked. There have been many people who have jumped straight in at the deep end and purchased very sophisticated and expensive machines, only to find that they don't understand how to use them. As a consequence, they either get sold or end up gathering dust in a cupboard having never had a sniff of a bit of silver.

Most detectorists begin either with a cheap basic model, or one costing little more than a couple of hundred pounds. With experience gained, they then usually trade up for something more upmarket.

If you're unlikely to spend much time out in the fields then an expensive machine may not be a good investment. Saying that, of course, there have been occasions when fortunate prospectors have ventured out for the first or second time and found something worth more than the machine they have just purchased!

This actually happened to me (Dave Stuckey). After having purchased my first machine in 1976 at a staggering cost of £80, I took it out for its first trial and within minutes I found a solid gold pocket watch. The scrap value of the gold more than covered the cost of the detector.

To explain the differences between the different types of metal detectors available it is necessary to go a little into their history and development. (If some of the terms used are unfamiliar please see the Glossary included later in this book).

The first types of hobby metal detectors to become readily available in Britain in the late 1960s were BFO (Beat Frequency Oscillation) models. These were very basic and thus cheap and easy to produce....and buy. Most had two basic controls: on/off-volume, and tune. Tune consisted of setting the detector to a faint ticking noise that could be heard continually from the speaker (or headphones). When a target was located this ticking noise would increase in frequency. They had single coil open search heads, which made pinpointing a find difficult. Depth was seldom more than a few inches. They had no discrimination and would pick up nails, silver paper and ring pulls as well as wanted targets such as coins. They were also subject to drift and needed retuning every few minutes (or sometimes less!). Despite their primitive nature, fields, commons and other sites were "virgin territory" and some good finds were made by their users.

Several American detector manufacturers produced more advanced models of BFOs in the 1970s, but other designs eventually overtook these early machines. To my knowledge none are available new today although some may still be found at car boot sales or in junk shops. However, they are more of a curiosity and collector's item than a usable piece of equipment.

The next model to become available in Britain in the early 1970s (although already established in the USA) worked on the IB/TR (Induction Balance/Transmit Receive) principle. These had two balanced coils in their search head, and would signal when a metal object disrupted their electromagnetic search field. Depth and pinpointing was a lot better than on earlier models. Although at first they did not have variable discrimination, some would naturally reject small pieces of iron or tiny fragments of silver paper. Most models, again, had two controls: on/off-volume, and tune. Tuning consisted of adjusting the control so that just a faint noise could be heard. When a target was located this would increase in volume. However, the problem remained of drift and the need for manual retuning when this occurred.

The problem was partially overcome with the development of push-button auto-retuning. This new control was a button that was held down while the detector was being tuned to threshold and then released. If the detector drifted from its pre-selected threshold point, all that was necessary was to push the button to bring it back to threshold.

Variable discrimination, which appeared at around this time, was another technological advance. Detectors of this type were usually referred to as: "TR/Discriminators". Here a rotary control could be used to set the detector's reject level to unwanted junk targets. However, the use of discrimination brought three disadvantages: it increased the detector's susceptibility to mineralised ground; it caused loss of depth; and it meant that wanted targets were sometimes rejected alongside unwanted ones.

The next development, in the mid-1970s, was the VLF/TR (Very Low Frequency/Transmit Receive). Detectors of this type had, in effect, two separate circuits. One circuit, the VLF was less prone to mineralisation and could provide good depth. However, it was all metal and could not reject junk items. The TR side could discriminate but had less depth. Thus searching was carried out in VLF mode, and once a target was located the detector was switched to TR to establish if it was junk or a wanted find. The problem with this system was the need to switch back and forth between the two modes.

The invention that overcame this was meter discrimination. In detectors of this type both circuits were working at once, the VLF all-metal side controlling the audio while the TR discrimination side worked the meter. Thus all registered targets came through on audio but the meter needle swung left for junk, right for wanted targets. Further developments included variable discrimination on the TR side, and an overlay of tone on the VLF side (low for junk, high for wanted targets). On some models of this type of detector it is possible to select meter or tone discrimination, or both. Such detectors are still in production today and have proved very effective in the hands of their adherents, particularly on finding good targets amongst high levels of junk contamination.

Most detectors today are of the "motion" type that made an appearance in the late 1970s to early 1980s. Such detectors can overcome ground effect while discriminating at the same time by means of continually and automatically auto-tuning. At first the sweep speed had to be very fast, but further developments slowed this down to normal. To register a target with a motion detector the search head has to be in movement. If the search head is held stationary over a target the auto-tune circuitry will simply cancel it out. However, many motion detectors have an all-metal, non-motion mode for pinpointing and, even without this, pinpointing can

Fitting a scuff cover protects the search coil.

Early TR models.

An early 1980s detector working on the meter discrimination principle.

Hip or chest mounting a detector can save arm fatigue.

11

be achieved quite easily as the coil movement required to register a target can be very slow.

During the mid to late 1980s computer technology began to be incorporated into metal detectors. In the same way that it is possible to programme a computer, it is now possible to programme certain detectors with a host of variables by means of an LCD screen and touch pad controls. These detectors come with basic "factory pre-set" programmes or you can devise and store a programme that you have developed yourself to cope with the conditions of a particular site. Some detectorists take readily to this new technology while others prefer simple "switch on and go" machines. Space does not allow us to go into further detail, but such detectors are - and will continue to be - fully covered in the hobby literature.

In parallel to the developments described above have been those involving PI (Pulse Induction) detectors, first available to the hobby market in Britain in the 1970s. PI detectors put out a very strong electromagnetic field that energises metal objects buried in the ground and creates eddy currents around them. This makes such targets easier to detect. The system is virtually free from ground effect and Pulse detectors can be capable of above-average depths. However, PI detectors are sensitive to iron and many have no ability to discriminate against this or other types of metal. This makes them difficult - or often impossible - to use on inland sites. Battery drain is also often higher than with normal detectors.

Most modern detectors can operate quite well on wet sand on beaches if they have a "Beach" mode but Pulse Induction units are considered to be the best for this kind of site. The greatest advantage they have over conventional detectors is their greater depth-seeking capabilities and the ability to ignore mineralisation. Many come in waterproof cases that make them ideal for searching a beach down at the water's edge and into the surf.

At present several new developments are underway, and PI machines are actually available that the manufacturers claim can be used on inland sites. Future evolutions in PI technology are worth following with interest.

With conventional detectors there is a broad spectrum of makes and models to choose from. There are relatively low-cost machines that are simple to operate and give reasonable performance. Following this, there are many models that are more sophisticated but cost several hundred pounds. These types of detectors offer very good performance, however, and are probably the most widely used.

Apart from the detectors mentioned above, there are also more specialised machines designed specifically for underwater detecting and hoard hunting. The hoard hunting machines tend to be simply a large box with a coil at each end, and are designed to look for large targets, such as pots of coins, buried at great depths. They are, generally, less efficient at locating smaller targets nearer the surface.

A final consideration is that of weight and balance. Some detectors are light enough for long hours of use by youngsters or by the infirm/elderly; others are heavy and are best suited for fit adults. One way to overcome weight problems is to buy a detector with a detachable control box that can be belt or chest mounted, saving fatigue on the arm. If possible, "try before you buy".

DETECTING ACCESSORIES

Search Coils

Metal detector search coils come in a variety of shapes and sizes from just a few inches across to a staggering 18in! Some are concentric (polo coils) and some are even "web" shaped. Several makes of detectors offer the ability to interchange from one type of coil to another.

Smaller coils tend to have better pinpointing capabilities and are easier to use on rough ground or on sites that are overgrown and where manoeuvrability is a problem.

Larger coils usually offer greater depth capabilities and can cover more ground in a shorter space of time. These coils quite often add considerable weight to a machine and are more suited to flatter ground surfaces.

Concentric coils put out a search field resembling an inverted cone. Size for size they usually go deeper and are often better when searching junk-infested sites. Double-D or wide scan coils, with their cylindrical search fields, normally provide better and faster ground coverage.

It is worth checking, before you make a purchase, that the type of detector you wish to buy can use interchangeable coils.

Examples of coil types.

Concentric.

"Web" shaped.

Widescan.

Eliptical.

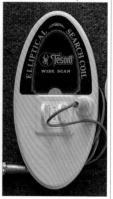

Sand scoops.

Headphones

Although most metal detectors have built-in speakers the use of headphones is recommended. These help to exclude outside noise that can mask any faint signals.

You will find that most metal detectors will generally accept any type of headphone from the portable CD player type right up to the largest size available. Most dealers also stock more expensive headphones that offer a "signal boost" facility. Again, it all boils down to what you are willing to spend. Headphones also help to keep your ears warm in the winter!

Digging Implements

It can't be stressed enough that the need for a good quality digging tool is essential in this hobby. Whether you opt for a simple trowel or a foot-assisted spade, you would be well advised to buy the strongest that you can afford. A cheap "tinny" trowel may be ideal for garden use but it won't last five minutes in the field. This is particularly important if you intend to detect in an area where the soil is of the heavy clay type. I have seen, and personally experienced, several occasions when digging implements have bent or broken completely while in use. This can really be a nuisance when a promising target has been located but the finder can't get to it due to implement failure. It may also be advisable to keep a spare tool somewhere near to hand - as I do.

Coin Probes

These are small "metal detectors" that are basically not much bigger than a pen and cigarette packet and fit easily into one's pocket. They are used only when a small target proves difficult to locate in a hole. The end of the probe is swept around the inside of the hole and a small LED lights up when it comes into close proximity with the coin or artefact.

Finds Bags

The worst thing you can do to a precious find is put it straight into your pocket along with all the other odds and ends that you have extracted from the ground. This will probably do more damage than centuries of agriculture and corrosion could ever do! Most detectorists wear waistband type pouches to carry any "good" finds. It is also recommended to carry a selection of self-sealing polythene bags in order to keep delicate finds separate from one another.

Basic Kit

Apart from the above, basic kit should include warm or waterproof clothing (depending on weather conditions), sturdy boots or Wellingtons, gloves, water, First Aid (sticking plasters), notepad and pencil and insect repellent. Also, either tell somebody where you are going or take along your mobile phone.

WHERE TO DETECT

Once you have bought, borrowed - or perhaps been lucky enough to have been given a metal detector as a Christmas or birthday gift - the next consideration concerns where you can use it.

The information in this chapter will, hopefully, provide you with some idea of the places worth searching, how to find them, and (most importantly) how to seek search permission.

Gardens

The first place you will obviously try is your own garden, if only to see how your detector works. If you live in a fairly modern house, you need to bear in mind that it was once a building site and you will, without a doubt, find a plethora of associated junk. By this I mean such things as nails, rusty metal, silver paper and cigarette cartons.

If you live in an older house, or know somebody who does and is willing to let you search their garden, you stand a reasonably good chance of making some interesting finds. Many people have lost rings or other items of jewellery in their own gardens. I was once asked to come and search the grounds of a large country house after the owner lost a valuable signet ring while playing badminton on the lawn.

Of course, the older the house the greater chance you have of finding something of historical interest. I once asked permission to search the grounds of a large house that dated back to the 17th century. On the lawn at the back of the house I unearthed a beautifully ornate trigger guard from a flintlock pistol, which was contemporary with the house. I gave this to the owner, who was very pleased with it.

If you are friendly with your local vicar then it is always worth trying to seek permission to search the grounds of the local rectory. Many tea parties would have been held in the gardens offering the chances of making some interesting finds.

Parks & Commons

If no restrictions on the use of metal detectors exist (check the local bye-laws) then parks and commons can offer a good chance of making some interesting finds. The age of the park, or common, will reflect the type of finds you can expect to make. Don't be put off by parkland that is surrounded by modern development; always remember that it was probably farmland at one time.

If there are any very old trees in the park then search around them; people will have sat or even picnicked under them at some time in the past. The large oak tree featured in Fig.3.1. is situated in a park surrounded by modern development. When I searched around it some years ago I found over 40 coins dating from George III to the modern day. Under another tree, some 50yd away, I found a large, chunky-linked gold bracelet worth several hundred pounds.

You will undoubtedly find masses of silver paper and pull-tabs but with careful use of discrimination, or "notch", much of this can be eliminated.

CM

Fig.3.2. This 5th century Roman gold solidus was found under bushes on a common. Scale 200%.

Commons are usually very old; many of them were used hundreds of years ago by villagers to graze livestock. In the past some spectacular finds have been made on them. One common in particular, near where I live, yielded a hoard of coins dating from the Civil War. When I searched it for the first time in 1986 I found the Roman gold coin shown in Fig.3.2. under some bushes! Again, it is always wise to check the by-laws before attempting to detect on commons.

Fig.3.3. Julian and friend Jeff work a ploughed hillside.

Ploughed Fields

Ploughed fields generally offer the greatest prospects for making good finds and are the most favoured sites for any detectorist. This is mainly because they are continuously being turned over, bringing new finds to the surface. The only drawback to these sites is the damage caused to many finds by farm machinery and long-term exposure to agrochemicals. This doesn't mean that you won't find anything in good condition, however. As an example take the Roman brooch in Fig.3.4., which was found on plough-soil and is in remarkably good condition.

To give you the basics of which ploughed fields are best to search would take a whole volume in itself. But you may improve your chances by doing plenty of research, which is covered in Chapter 4.

Generally, any fields close to an ancient settlement will probably yield finds associated with it. Detecting near any Roman sites, for example, will almost certainly yield things such as Roman coins, brooches, tools and other implements.

Fig.3.4. Undamaged "trumpet" type brooch found on a Roman site.

Fig.3.5.
Fields near old churches, such as this one, often yield many finds and are always worth searching.

Fig.3.6.
"White metal" medallion to commemorate the Silver Anniversary of the reign of George V.

Many Roman settlements or habitations were established on sites that already existed prior to the Roman occupation; this means that you could even find coins and artefacts that pre-date the Roman period. These could include coins and artefacts from both the Iron Age and, possibly, the Bronze Age.

It is also worth searching fields close to Roman roads. These often yield occasional Roman finds, or even hoards of coins deposited by merchants who feared attacks by bandits.

Fields close to old villages will always yield finds lost by earlier inhabitants. Prior to the Second World War, the vast majority of villagers were employed on the land and the variety of things that many of them lost will amaze you.

Fig.3.7. Crotal bells are common finds.

You will find an abundance of buttons, thimbles (for repairing clothes in the field), coins, jewellery, cutlery, pipe tampers, lead bale seals, medallions (see Fig.3.6.) and, of course, harness fittings. Another common find is the crotal bell (Fig.3.7.). These usually date from around the 17th century and were used on harnesses, etc.

Fields close or adjacent to old churches are always worth searching. Since medieval times, the church would have been the centre of activity in any village. Most modern towns and villages started out as small settlements around a church and many losses would have been made near them.

The most common find will be what is often regarded as the bane of all detectorists - the shotgun cartridge cap. Don't be tempted to throw these back into the ground, however, as you'll only end up digging them up again at some time in the future.

Fig.3.8. *Dave Stuckey searching a meadow with an old Arado 120B.*

Pasture

Pasture fields probably have only two distinct advantages over ploughed fields. Firstly, unlike ploughed fields, you can search them at almost any time of the year (providing they don't contain livestock). Secondly, due to less soil disturbance, the finds often come up in better condition.

Fig.3.9. *These predecimal coins were all found on pasture land.*

The only real drawback to these sites, however, is that once they have been detected thoroughly they don't yield much afterwards - unless the farmer disturbs the soil, that is.

It is worth remembering that many pasture fields have at some time in their history been ploughed for one reason or another, leaving the strata in disarray. Don't be too surprised, therefore, if you find something Roman at, say, 2in depth and then unearth a Victorian penny at 10in!

Following the same guidelines as for ploughed fields will once again improve your chances of making good finds.

Footpaths & Woodland

Detecting on footpaths and in woodland is one way of keeping yourself active during the "dormant" season (by that we mean during the summer months when ploughed fields are inaccessible due to crop growth).

Footpaths, particularly well-used ones, will undoubtedly yield coins and other objects lost by ramblers. Larger trackways and footpaths, which date back centuries, could yield more interesting finds.

One very old trackway, which we searched some years ago, yielded a gold half-sovereign, Victorian coins, and a silver shoe buckle dating from the 18th century.

Sadly, you will also find considerable amounts of modern refuse, such as: pull-tabs, silver paper and even drink cans, etc.

Stiles also present good opportunities for finds as many ramblers will have had to struggle over them, losing coins and other objects in the process. One stile, which was searched just a couple of years ago, yielded a pair of halfcrowns (Fig.3.10). As I held them in my hand I imagined some rambler, four decades earlier, arriving at a country pub hoping for a thirst-quenching pint only to find that he had lost his beer money!

Woodland detecting can be frustrating at times, mainly due to the vast amount of shotgun caps that they are invariably littered with. To add to this, we have yet to find a piece of woodland that hasn't seen action by the Home Guard, or regular soldiers on Second World War manoeuvres! If you ever attempt to search woodland you will, without a doubt, find copious amounts of expended ammunition of one type or another.

But don't be deterred - woodlands can reveal surprising finds. Despite the bullets and shotgun cartridges you will also find coins, jewellery and just about anything mankind is capable of losing while out on a woodland walk. Fig.3.12. shows a beautiful bronze medallion commemorating the visit to Canada in 1939 by George VI and Queen Elizabeth. This was found in an area of woodland where numbers of military badges and coins were also being found. No doubt it was lost by some patriotic squaddie during the war.

Fig.3.10. Two halfcrowns found together beside a stile. They were perhaps lost by a rambler of yesteryear.

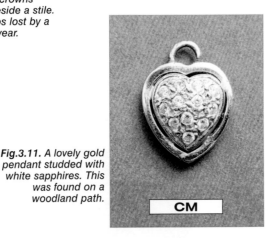

Fig.3.11. A lovely gold pendant studded with white sapphires. This was found on a woodland path.

Fig.3.12. Bronze medallion commemorating the visit to Canada in 1939 by George VI and Queen Elizabeth.

Rivers

Rivers have been the focus of activity throughout history, either as a means of transport or navigation, or simply as a convenient water supply. Large rivers, such as the Thames, have seen so much activity over the past three or four thousand years that it would be impossible to estimate what treasures lay hidden in their silt.

The Thames itself has yielded vast amounts of finds over recent decades, and most of those simply from the mud on its banks! These treasures include coins, jewellery, weapons, bottles and vases, and even a fabulous prehistoric bronze shield from the Iron Age, which is now in the British Museum.

Fields, or open spaces, beside rivers are always worth searching. You may stumble across a previously unknown ancient settlement, which could yield interesting finds, or simply losses made by travellers following its course over past centuries.

Fig.3.13. This George III "Hanoverian" shilling was found on the banks of a river.

If you can find areas where the banks have eroded, try these as well - you never know what has been washed out of the soil.

Fords, particularly if they are on a Roman road alignment, could also be productive. The Romans often cast coins or other votive objects into the waters for good luck on their journeys. It is probably a good idea to try these sites during the summer months when the water levels are at their lowest, or completely dried up.

Detecting on the River Thames does require a permit, which has to be acquired from the Port of London Authority.

Fig.3.14. Old windmill sites are also worth investigating. This is a selection of finds made on one visit to the site of a long-disappeared windmill in Hertfordshire.

Windmill Sites

In the past the British landscape was dotted with countless numbers of windmills - some dating back to medieval times. However, over the last couple of centuries their number has dwindled dramatically. Few survive in restored condition and many of the remaining ones left standing have been converted into modern dwellings.

If you can locate the sites of demolished windmills you stand a very good chance of making a large number of excellent finds. These include coins, buckles, buttons, rings, seals, bells and many other types of objects. Fig.3.14. shows a selection of finds that I made on a single visit to the site of one particular windmill near my home in Hertfordshire.

A windmill had existed on the site since medieval times but was demolished without any trace at the end of the 19th century. Fig.3.15. shows a superb sword chape, dating from the 15th century, which I found on the site.

Most landowners will know of any windmill sites that exist on their land, or you may find the relevant information at your local library or Public Records Office.

Fig.3.15. 15th century "Castellated" sword chape.

CM

Beaches

If you are fortunate enough to live near the coast you will almost certainly want to try out your detector on the beaches. Not only are they available to detect on all year round, but they can also be very fruitful in finds, particularly if you follow a few basic tips.

Most of the jewellery losses made on beaches occur when people go swimming in the sea - their fingers contract, due to the cold temperatures, making rings very loose and liable to slip off.

Bathers also tend to lose necklaces when they do the breaststroke. Forgetting that they are wearing them, they push their arms forward and outward - breaking the chains in the process.

The tip here is that the best time to detect on a beach is when the tide is out. Start at the water's edge then work backwards and forwards as the tide returns. Tide charts are available in many seaside resorts and are well worth obtaining.

There is no best time of the year for beach detecting, although winter storms can churn up beaches well enough to reveal long-lost jewellery and coins.

I remember on one occasion, when I was detecting on a beach near Portsmouth, being approached by a very distraught couple who begged my assistance. The lady, who had been swimming in the sea, came ashore to discover that she had lost both her wedding and engagement rings! Both rings had a joint value of several thousand pounds. Unfortunately, she had been swimming too far out and the tide didn't recede enough for me to recover them. This was indeed a sad tale, and one that left me feeling distraught about being unable to assist. I would have given anything to have been able to relieve the lady's anguish by retrieving the rings.

If beach detecting is going to be high on your list of search sites, then you must ensure that any machine you choose to buy is capable of searching beaches; some types of detector do not work well on wet sand.

Where Not To Search

The previous suggestions are just a small selection of the kind of sites that are worth investigating with a metal detector. There are, of course, some areas where detecting is usually forbidden. These include the following.

Scheduled Sites

Scheduled sites are normally archaeological areas that have yet to be excavated. Along with Ancient Monuments and ruins, they are protected by law and should be avoided. Your local library should keep a copy of the Scheduled Monuments Record. If not, then the Archaeological Department of your County Council should be able to advise you.

Most landowners know whether any of these sites exist on their land and will deny permission to detect on them.

Sites Of Special Scientific Interest

SSSIs (or Sites of Special Scientific Interest) are usually conservation areas that have been set up in order to protect wildlife and plant life. Again, there are laws protecting these sites and detecting is usually out of the question.

RESEARCHING POTENTIAL SITES

To find the most promising and, hopefully, most productive sites research is necessary. After you have located potentially good sites the next step, of course, is to gain permission to search them.

Naturally, the fact that you have intensively researched an interesting looking area does not mean to say that when you approach the landowner you will gain search permission. However, if you take along the results of your research when you knock on the door it can help weigh the odds in your favour. Even if you should receive a rejection your hours or days of research will be by no means wasted. The exercise will have increased your knowledge of local history, and - from the experience gained - will make the next research project that much easier.

Much of the documentary site research that you will conduct is the paper equivalent of fieldwalking. Where appropriate in this chapter, therefore, helpful hints and links to fieldwalking are included to expand the topic. The majority of these additions are based on our own experiences.

There are a number of sources of invaluable information to assist you in finding promising areas worth searching. The following suggestions are the recognised areas of research that have been adopted by The Pastfinders (of which Dave and myself are members). We have found them all highly effective.

Modern Ordnance Survey & Older Maps

Maps such as these can be used to examine field shapes and road patterns, and many historical sites are shown on them as well. A lane or track, for example, which has a semi-circular or basically circular route may indicate an enclosed settlement that the road or track went around. Roman roads will deviate from straight alignment normally only as topographical requirements dictate, but we once discovered a former Roman temple opposite a small un-required deviation. In all probability the deviation may not have been temple linked, but it shows how looking for the slightly out of the ordinary features can bring rewards.

Other places worthy of investigation are crossings of ancient trackways or marked Roman roads. The most ideal modern maps for this purpose are the Ordnance Survey "Explorer" and "Pathfinder" ranges, both at a scale of 1:25000 (2.5 inches to the mile).

Please note that the green coloured cover Pathfinder range of maps is unfortunately no longer published, but copies are still worth obtaining if you encounter them second hand.

Many moated sites, coin hoard find spots, and old windmills are also marked on such maps. The surrounding areas to these should reveal finds. Look for isolated churches that may indicate a village abandoned due to the plague. Landowners re-designing their estates or areas around these churches could reveal traces of the original settlement. Small clusters of isolated ponds could be the carp ponds from a long-vanished monastic building or early manor house.

Isolated large trees, copses and field corners near farms and villages are also easily spotted on such maps. Such places are where picnickers and farm labourers of the past may have taken their work breaks. Also look for rivers and spring sites that could have supplied water to a nearby settlement.

A number of websites are available that allow you to look at old maps of your area. These can be invaluable for locating lost footpaths, and many show windmill sites etc that do not always feature on more modern versions.

Tythe maps are an excellent source to study field shapes prior to Enclosure actions of the last 400 years. The majority of these often tiny individually held tythe plots have long since disappeared.

Study the contour lines on maps to locate rises in land; in flattish countryside hills and slopes have attracted people for many millennia.

Libraries, Museums, County Halls, Churches & Public Records Offices

Many libraries stock a good selection of general, and more local books and records; some even stock old aerial photographs. These are mostly 1940s and 50s RAF exposures, but some sources also have Luftwaffe photographs as the German air force extensively photographed the UK during reconnaissance in the early years of the Second World War.

Aerial photographs are as good as maps, with the added bonus that they will show you crop markings and other potential sites not usually detailed on maps. However, one should always use caution, as there are quite a number of pitfalls in site interpretation from aerial photographs. One example we have encountered is Second World War ploughed out bomb craters and search light battery sites that can look like Bronze Age and later ring ditches. But if you are interested in searching for World War Two artefacts, then perhaps it's just as appropriate to consider the reverse.

Yet another factor that can easily fool the inexperienced and experienced alike is a geological fault that can occur on some chalk hills. This is where soil slips from the peak downwards; in severe cases this can form a band of rings around the hilltop. Natural occurrences such as these are easily mistaken for ditches associated with Bronze Age or other settlements (please refer to Chapter 6 for further examples).

Regarding aerial photographs there are again a number of very good web sites that cover most of the UK, and the potential of these is phenomenal. Your local museum will also be a rich source of information. Many museums now have excellent relationships with detectorists, and quite a few stock an excellent display of published books and pamphlets on local history. Most museum staff will only be too pleased to put you in contact with county archaeological departments, Finds Liaison Officers, and local historians. Some museums also stock records of archaeological digs and investigations going back decades, many beyond living memory.

Public Records are a rich source of research information and hold a wide variety of records that may be useful to you, such as Tithe Maps.

Churches may be the centre of a local magazine publication that contains many interesting facts and details about villages etc. Some churches will also house the parish records, which can be viewed by appointment. Parish records for the years 1939-50 can be very rewarding to look at. The principle reason for this is that huge areas of pasture and virtually any available plot of land were put to the plough or the spade to increase this country's food production. Consequently more artefacts, coin hoards and sites of historical interest were uncovered than ever before, and many were reported in the local parish records.

County Halls can also sometimes be good sources of information. One we know of houses a comprehensive record of all known crop markings in that county. All of these sources are extremely good for research and gathering information; how you use it will undoubtedly be reflected in your future finds rate.

Local Newspapers

These can be very good for articles on local archaeological discoveries and more often than not individual metal detector finds. If you are new to the hobby the article may include a contact name or group who you could approach.

Local newspapers are as always a good way of getting to know what's going on in your area. Such local newspapers also carry articles and notices in relation to new developments such as roadways and housing estates. Both sites could be worth obtaining permission to search on. Should these new developments be in areas of historic interest, find out if there is any rescue work going on. Perhaps the group doing this would welcome somebody to help metal detect the spoil. One such notice relating to a new bypass in our district led us to gain permission to briefly search the cuttings made along its route. Most of these were dug into chalk, so any interesting soil disturbances showed up clearly. When we arrived we noticed a series of circular clay patches full of round pebbles. Numerous worked flints and cores were scattered about and inside these circular patches. We later discussed the finds and concluded that most likely the new road cutting had truncated some prehistoric flint mines. Later when glaciation occurred after their excavation, these holes became filled with clays and thousands of pebbles as the huge ice mass ground and slid its way over them.

Talking To Local People

It's a very good idea to talk to local people about your new hobby. For instance, somebody might remember their grandfather ploughing up some coins before the war. Developing a network of people, who are aware of your interests, will help your research. Public houses can be very good nodal points for meeting people and asking who owns such and such bit of land etc.

If you are lucky people will start to contact you, telling you when a certain field is ploughed, or that Mr. Smith has a collection of coins from his garden etc. Agricultural workers in many cases will be aware of features that are of interest to you, due to the nature of their work.

Is there a local historian who might be able to assist you? You might even hear of other local detectorists, who - if they care about the hobby - will I am sure help you with the early stages of research. (Although don't expect that they will always tell you where all their wonderful Roman *denarii* came from; sites like these you will have to find for yourself).

Finally, if these people are helping you and they are in turn interested in history etc, do inform them of what you are finding. Your finds rate in a certain area may be directly connected to a bit of information given to you. We share knowledge related to this hobby with a wide variety of people and organizations including schools, other detecting groups, history societies and museums. In turn, a reciprocal flow of information comes back.

Fieldwalking

Fieldwalking is the final part of site research (dealt with in greater detail in Chapter 6). There are no hard and fast rules to the order in which you conduct site research and permission seeking. When you have conducted all your documentary research, it could be that the areas you are interested in have rights of way across them. If this is the case you could have a look at a site before asking permission. Who knows, you may even spot a scatter of pottery or oyster shells. Even better, you might meet the landowner or a farm worker and outline your interests to him.

It is important to remember that despite land having public rights of way across it, until you have secured the landowner's permission you must not use your detector to search.

GAINING
SEARCH PERMISSION

Seeking Permission To Metal Detect Or Fieldwalk

As with so many aspects of this hobby there are no hard and fast rules to gaining search permission. However, the advice that follows is based on many years of experience and should produce good results.

Our suggestions are by no means exhaustive and you may, on occasion, simply need to use common sense and tailor your research and permission seeking to suit particular areas, and individual landowners. For example, carrying out a little background investigation into the interests of a landowner will give you a conversation point should the opportunity arise.

When making an approach to a landowner, we have found it best to be casually but well dressed, and make our visit at an appropriate time (eg not during Sunday lunch, or at a very busy time of year such as the harvest).

If there is a group of you asking, it's perhaps best if you send one person to the front door, as this is less of an imposition. Some landowners will say "No!" and may be rather offhand; you must be prepared for this. Should you receive an abrupt negative, simply say something along the lines of

"Oh well that's a shame, but thank you for your time anyway". Never argue the point, even though it may be very frustrating; always be courteous. You can depart by offering to leave your telephone number should they have a change of mind or circumstances for refusal change.

The Pastfinders are very lucky in that they all combine effort to produce articles for **Treasure Hunting** magazine. One of the greatest benefits to this is that we take past copies with us when we seek permission. We have found many landowners are very impressed, and on several occasions we have turned "No" into "Well, perhaps it would be rather interesting". Not everyone writes articles, therefore as an alternative option why not take along a small neat portfolio of documents relating to your research.

At all times try to put forward the advantages of having a detectorist operating on the land, and show a passion for your hobby. We have found many landowners are fascinated by the history of the land they farm. Sometimes you may hear "Well, I already have several people who come once or twice a year and detect all my land". This is sometimes a simple refusal with an excuse; but it is not always an unredeemable situation. Try and find out who these other detectorists are. Perhaps they are members of an established club and you could even join it.

As members of The Pastfinders group we are fortunate enough to have quite a few sites to search. When we are attempting to obtain permission for other areas and such is granted, we always ask that the landowner does not deny permission to other detectorists just because he has given it to us. We just do not have enough time to cover all of our sites, and would hate to think that we are tying up large areas of land when other detectorists might be having trouble finding somewhere to search.

Above all, never be downhearted when you receive a refusal. Look forward to all the permissions that your hard work will eventually obtain for you.

Having researched the history of the land previously, you might well be able to inform the landowner of important historical happenings or previous finds that he/she was not aware of. In many cases you will find that this can be reciprocal, and the landowner can also tell you about past finds. This will be of great assistance in helping you to pinpoint productive search areas. A wonderful example of this occurred just as I was writing this chapter. It involved the landowner of an area where we have just secured permission to search. Research into this area had shown that there had been an extensive Roman settlement. We therefore wondered if, and where, there might have been an associated temple. Incredibly, the landowner revealed to us that when his father's house had been built, the workmen had found a number of Roman oil lamps and curious tripod-legged incense burners. Sadly, this means that the temple itself is most likely now beneath a modern building. However, arable fields surround this modern house and hopefully there will be some very interesting finds to be made in them.

It is necessary at this point to address an important issue. When you seek permission or obtain it please make sure that it is indeed the landowner you are dealing with. A tenant farmer might be very obliging but cannot legally grant search permission without the landowner's consent. This is worth bearing in mind, as you could well get the tenant into trouble as well as yourself, and all your previous research might be wasted.

Once ownership is established and you have been successful in obtaining permission from the actual landowner, it is a good idea to ask if there are any Scheduled Sites on his/her land. You must not detect on these without specific permission from the relevant authorities, and only in special cases is this likely to be granted. If the landowner is uncertain, then you will need to establish the possible Scheduled status of any sites yourself. Your local museum is one of the ways that you can confirm site status if you are uncertain or, alternatively, you could contact English Heritage.

One other consideration is required, and that is to know the exact extent of the land for which you have gained permission. It is a good idea to ask the owner to outline his/her property boundaries on a map. This will avoid any accidental trespassing on adjacent land on which you might want to gain search permission later.

When permission is granted you need to bear in mind that other people's pastimes may also be in operation on the land concerned. Ask the landowner if there are regular shoots etc on his land as you would be most unpopular if you appeared round a hedge into the middle of a pheasant drive.

At this stage it is also a good idea to discuss potential finds with the landowner. Remember, too, that if you are asking on behalf of other individuals make sure the landowner is aware of numbers involved. This is where some landowners may set conditions such as "I don't mind one or two but seven is a little too many". This, of course, is their prerogative.

Concerning any finds that we make, we normally operate a policy of giving the landowner first choice of anything we uncover. We show the person concerned all of our finds, including shotgun caps, lead dross etc. In the case of a really valuable find we normally abide by a 50:50 share with the landowner.

It is a good idea to discuss and establish a finds sharing agreement at the earliest opportunity with a written and signed contract. There have recently been a number of serious "fallings out" and legal disputes where these agreements have been neglected. We are fortunate in having many sites to search, and a number of very good friendships have developed between us and the landowners concerned.

It is important to maintain these good relationships. As a thank you, The Pastfinders present many landowners with a bottle of whisky or red wine at Christmas along with a card. Without the co-operation of landowners our hobby would not be able to exist in its present form.

We have found that a picture frame, containing a mounted selection of coins and artefacts from a farm, is always most gratefully received by the landowner and provides a good talking point for guests who visit.

One other service that can be offered is a free jewellery or agricultural implement part recovery service. We have found this to be highly successful. In offering it to friends and family of the landowner we often gain permission for even more land to search.

When you have got permission to detect upon what might be your very first site always remember that you now represent the thousands of other detectorists in the UK. Should you encounter inquisitive persons such as walkers etc, always take the time to inform them what you are doing. Show them the detector, how it operates, and any finds you have made. The more people who are made aware of our good work and efforts, the better. We have often found that stopping for a chat, or passing on a few Georgian coppers to such people often results in such suggestions as "Would you like to detect my garden?" or "I own an acre or so down the road, perhaps you would like to try your luck on that sometime?"

If you are polite and informative to the public, you might one day just be lucky and get to detect on a small seemingly insignificant patch of land that happens to have a hoard of coins on it!

Remove all rubbish that you find and dispose of it thoughtfully. Once you have permission to detect on one farm, you can quote this fact when you seek permission for different areas.

From this stage its up to you, but we wish you the very best of luck and success.

FIELDWALKING & WHAT TO LOOK FOR

As mentioned in Chapter 4, perhaps you may have already taken the opportunity to conduct some limited fieldwalking (along public rights of way such as footpaths, bridleways etc) prior to asking search permission from the landowner concerned. The hints and suggestions in this chapter are relevant to all types of fieldwalking, but perhaps best suited to that done once search permission has been obtained. In such circumstances - crop conditions permitting - you are unrestricted and can wander at will.

However, when you fieldwalk, what exactly are you looking for? That could be dependent upon your particular interests, and could extend from the Bronze Age to hunting Second World War relics.

The term "fieldwalking" largely applies to agricultural land subject to ploughing, but at the same time can take in a wide variety of other terrains such as chalk pits, riverbanks, and wooded areas. Woodlands, for example, cover many medieval moated sites, and depending on the season when your search takes place many factors dealt with below could be applicable. If you discover a site of interest, and there is an adjacent river, check the exposed banks; often stones, pottery and other clues can be evident.

A small river we checked recently had occupation evidence clearly visible in its banks as well as huge worked stones lying on its bed. This certainly stimu-lated our interest in this newly fieldwalked area, and resulted in some very interesting finds being made.

While on the subject of rivers, study the banks for erosion on both sides as this might indicate an ancient crossing place. Often there can be traces of the trackways in the same areas, particularly if the river course is surrounded by un-ploughed ancient meadowland.

However a field surface that has been deep ploughed, harrowed or rolled flat, will perhaps reveal the most to the keen fieldwalker, hopefully revealing the potential of metallic finds that may be associated with your discoveries.

Such surfaces may evidence pottery shards, stone scatters, odd coins, different soil colouration, tile, oyster shells and a whole host of indicators to past ancient activities.

This chapter will therefore deal for the most part with ploughed fields, but will also cover other pertinent details such as animal activity etc that we think may be helpful. Basically, when fieldwalking it can be helpful to look for features that are noticeably out of the ordinary context of your search area. For a simple example, in dark soiled fenland fields, which can be virtually stoneless, a large scatter of stones would be unusual and could indicate a building from some period. A large scatter of flints on the top of a

Fig.6.1. A stone scatter such as this may indicate a Roman building.

Hertfordshire chalk slope is almost always a geological occurrence. A patch of dark soil in a light coloured soil field, or a patch of light soil in a dark coloured field, would both be of interest.

However, never be too dismissive or too confident; many a Roman villa has probably been missed over the years by over-confidence, and not bothering to closely check something out.

We have found that for searching large fields a powerful pair of binoculars can be of great assistance.

If you are lucky, perhaps your earlier research has revealed a series of rectangular crop markings and you already have a firm objective to set out and look at these. There follows below a list of factors that we would pay attention to, or most certainly be looking at, to see if we could establish such a presence.

Fieldwalking is a marvellous section of research and can be extremely fulfilling. As you become more and more experienced you will find that you automatically make calculations that can make you smile as they lead you to find such a thing as a Roman farmstead. The presence of this may have been quite obvious to you, but never grow too confident for in your enthusiasm to reach this site you may have walked too fast and missed a smaller ancient farm right beneath your feet. Take your time and you should hopefully be rewarded with finds.

Experience can tell here, but it is often the metal detector that determines the final classification. For example, you have been out fieldwalking and in one area have found some grey coloured pottery fragments, one or two glossy red decorated shards, several oysters and some tiles, and a very lucky eyes only find of a hammered coin.

This means that the site you have found could be:-
1. A medieval site with some traces of earlier Roman habitation.
2. A Roman site with a casual hammered coin loss as the building materials were robbed out at a later stage.
3. A midden showing Roman, Saxon and medieval pottery.
4. A place where somebody has dumped soil from another district.

Perhaps a later metal detector search results in the recovery of over 50 late 3rd and 4th century coins, spread over this area and further afield. Here the metal detector has assisted in identification of the fieldwalking-discovered site, as to being a late Roman settlement.

Fieldwalking on its own is never easy or even conclusive, but it is always very enjoyable.

If you are fieldwalking remember to close any gates that you have opened, be aware of any livestock present, and obey any legal signs displayed. You should also take every precaution not to disturb breeding birds or wild life. As a basic rule any wild bird that ventures unnaturally close to you means that in all probability you are very near its eggs or newly hatched young. So walk carefully away. Fortunately, most fieldwalking occurs well after most of the breeding season has been completed. The following factors relating to settlement/habitation sites do not normally occur on their own, although it is likely, in some cases, that you will notice only one initially.

As you become more experienced, when you investigate a stone scatter related to a settlement etc, you will notice tiles, pottery and shells as well. Gaining fieldwalking experience will illustrate to you just how interrelated some of the features listed below can be.

Hopefully, these categories will help you to analyse just what it is you are looking at, as well as gaining the knowledge to dismiss the many misleading features and their causes. While it does not aim to provide all the answers, it is hoped it will stimulate your own path of research.

Stone & Tile Scatters

Concentrated stone and/or tile scatters can be a very good indication that the area you are fieldwalking has seen past human activity. It is a high possibility that almost every field you check will have one or two visible fragments of tile here and there. Some of these result from the medieval/Tudor and later habit of spreading tile over fields to assist with drainage. Hopefully, after closer examination they will be discovered in association with some of the other factors outlined below. Some stone scatters can be visible for many miles dependent on the type and colour of the soil they are present on. Stone scatters can represent debris from medieval churches, Roman villas, and a whole host of other buildings and reasons.

More recent buildings tend to be associated with brick and slate or tile scatters, and glazed pottery. Sometimes a stone and tile scatter can be related to a Roman building that is some distance away. We

Fig.6.2. Roof and floor tiles can be found around an ancient site in large numbers.

know of one example where the building materials from a villa were robbed out. The medieval or later builders deposited these materials by the side of a road. We thought we had a good site but when no Roman coins appeared we were therefore surprised to say the least. Several hundred feet away there was a noticeable dark soil patch with some cobblestones; when detected over this revealed the villa and many of its coins and artefacts.

This relocation and re-use of building materials is also reflected in the fabric of some older churches and walls. Frequently in Norman and later times the local Roman villa site was seen as a cost saving source of building materials. You can often spot Roman tiles and fragments of Saxon gravestone etc re-used in many old church towers. A really spectacular example of this is St. Albans Abbey adjacent to the site of the Roman town of *Verulamium*; there are literally thousands of Roman tiles visible in its construction. There is therefore a good chance that several stray pieces of Roman tile in a church tower could represent a wealthy Roman villa nearby.

Any isolated scatters of large rounded cobblestones are also of great interest, as they can indicate the ploughed-out floors of ancient dwellings. Scatters of approximately 1in cubes of tile are other signs of a possible Roman building site. These tile cubes are called *tesserae*, and should you find them made from limestone, fired clay, granite etc, you may just have located a plough damaged mosaic site.

Huge areas of smallish rounded pebbles and gravels are, however, most likely to be glacial deposits. Beware of other geological stone scatters, such as the previously mentioned flint nodules on the crest of chalk slopes; but in all cases, if you are suspicious check it out. World War Two bombs caused large craters when they exploded, which today can still be defined in some instances by variations in soil colour or large stones being present on a field surface. These stones having been brought up from much deeper stratum by the explosion involved.

Despite all the potential pitfalls, however, when you discover your first Roman or medieval building you will notice many of the other facts listed below. One thing is for certain: having acquired the experience you will never forget it. This experience will build and develop; believe us when we say that in years to come somebody will show you something and you will know exactly what it is. Often at that point you will marvel at the fact that just five years ago you wouldn't have had a clue. You will soon build up an idea of what stones etc are natural to your area, and this in turn will enable you to spot any abnormality all the sooner.

One site was indicated to us by the presence of fractured but worked igneous rock fragments of a type not glacially transported nor naturally occurring in the district. They were, in fact, fragments of Andernach lava used by the Romans for making querns. This triggered a fieldwalk of the surrounding area, whereupon a very dispersed Roman settlement was located. Quite incredibly, a tiny piece of stone had given us yet another Roman site to search.

When fieldwalking on Roman sites it is well worth checking on both sides of any tile fragments you find; it may be a surprise at just how many animal paw and other impressions that are evident. The most amusing example of one of these I have seen, are the tiny feet impressions of a mouse, followed by the paw imprints of a cat!

Scatters Of Pottery Shards

Pottery fragments can be found on a wide variety of settlement sites from those of a Victorian manor house to a Saxon settlement. As a simple general rule, grey brown and black gritty fabrics are old or ancient, while most glazed fragments tend to

Fig.6.3. Fragments of Roman samian ware pottery.

be medieval to modern. Although some late Roman pottery imported from the Rhineland, and to a smaller extent produced in Britain, was also glazed in this country it is rare to find either. Grey, sandy, gritty pottery fragments tend to be referred to as "grey ware". It can be extremely difficult to distinguish medieval grey ware from that which is Roman. Therefore you should not definitely identify a site type without taking other factors into consideration.

Just to confuse the issue, some sites have pottery fragments from all ages over their surfaces. In this instance the metal detector will certainly assist in giving an idea of who mainly resided on such sites. Some Victorian bottle dumps have been ploughed out creating huge scatters of pottery often in association with other domestic items such as bones and ash. Apart from modern fragments, which should on the whole be easy to determine, there are also some ancient "exotics" that the inexperienced fieldwalker may accidentally dismiss as being modern. The chief one of these is called samian ware, a bright red glossy fabric often slip or mould decorated. Other Roman types may be black, grey or white with coloured slip decoration equally confusing to the beginner.

Some ancient pottery has crushed shell, calcite or egg fragments in its make up, and these inclusions can be seen as flecks of white within the matrix. Some pale fragments of Roman *mortaria* have small igneous rock or quartz inclusions, making them appear speckled. These inclusions, known as *trituration* grits were pressed into the surface before firing to assist with grinding grain.

Other ancient pottery sparkles in the light due to its clay having a high level of mica in it, or having been dusted over with it as a pre-firing preparation. Any dark pottery with wheel or star-shaped circular punch mark decorations is most likely to be Saxon.

Bronze Age pottery, on the whole, is quite rare to find on any surface, as due to its poor firing it can be very delicate. Neolithic pottery too is quite an uncommon find, but can be observed scattered over areas of known flint working.

Pottery fragments in their own right can make up an interesting collection; we have some Roman examples that still have the potter's finger print and nail marks on them.

Should you find patches of stones, burnt soil, and numerous pottery fragments a good chance is that you may have discovered a ploughed out kiln site. Be on the look out for misshapen and distorted fired fragments called "wasters"; these are classic indications that you are near to a kiln. Once you become reasonably proficient at pottery identification this will become yet another tool for you to use in possible site identification and therefore age.

Finding Freshwater & Marine Shells

Shells, particularly oysters, are always a good indication for areas of settlement. They were regarded as part of the staple diet of the poor from Roman times until relatively recently. If you find shells in an area they may well be the only obvious remaining indication of a midden. Middens are where household and general domestic rubbish have been deposited; sometimes they are near or actually in settlements, at other times they are situated a fair distance away or dispersed widely as a result of manuring. Another source of shell deposits are dried up lakes and meres. However, despite these being natural deposits of shells you should not dismiss them. Such areas have often encouraged adjacent settlement and often extensive fishing and hunting from Neolithic times to when the water was drained etc.

As a discarded oyster ages it becomes paler, with more recent shells being grey brown in colour. Roman and medieval examples are normally snow white with a slight flakiness to their texture. The good thing about ancient oysters is that being white they show up very clearly against most soil backgrounds, and this is even more the case after rainfall.

Small types of white coloured, flattish snail shells found in soil may well indicate that your search area was a lot wetter in past centuries. Such snail shells can also be evident in the ploughed out ditches of settlements and barrows. Other shells to look out for, particularly inland, are fresh water and marine mussels as well as cockles, clams, periwinkles and limpets. It is said that the Romans introduced the edible snail to this country and, true enough, we have several Roman sites where these still abound. If you have never seen one of these creatures before, you will be surprised at their huge size compared to the garden snail. After this shock spare a thought for the possible Roman site that could well be nearby. Where these creatures have once existed and died off you can find many shells; however, like oysters the really old ancient examples are very pale, some being almost white. Texturally, these snail shells can also become quite flaky when extremely old. Look for evidence of these shells in new road cuttings, or along ancient sunken lane verges.

Areas Of Light & Dark Soil Colouration

As mentioned before, a dark area in a light soiled field, and a light area in a dark soiled field could both be indicators of some level of soil disturbance or alteration. Sometimes you may experience both together. The most dramatic of these can be where strip lynchets have later been ploughed away. In

Fig.6.4. Darkened soil may indicate an occupation site.

really defined examples, the whole field can be covered in 5-15ft thick bands of alternating light and dark soil. Such patches are fairly obvious in their appearance. However, they are not always indicative of human habitation or settlement sites. Only closer examination of their surfaces will reveal this. For example, a recently laid pipeline, or an ancient hedge that has been grubbed out may cause a single dark line in the soil. In the case of the latter, a glance at the tythe map for the area would help in confirming this.

Landowners who have filled in moated sites on their land have created another example of soil variation. Much of this happened in the 1960s when scheduling was not so defined, and it was undertaken to ease ploughing of a field.

Sometimes the positions of ancient kilns can appear as individual small light and dark patches. On occasion, in association with kilns, you will find areas of burned soil that have been ploughed up. Most soil varieties that have been subjected to extreme heat normally turn brick red to orange, with areas of darker carbonisation. Dependent on what the kiln's use was, it may be surrounded by a dense mass of pottery fragments or nodules of iron slag. Some grain drying kilns have even been found in association with large amounts of carbonised cereal grains. These burned grain deposits have been at

Fig.6.5. The top layers in this chalk pit revealed much evidence of Roman occupation.

some depth; normally any burned grains found above 2ft depth originate from the now banned stubble burning days.

Normally, old habitation sites that have richer, darker soils will be associated with more defined darker crop growth than surrounding areas. Some dark patches have also resulted from neglected chalk pits that fill up with topsoil and organic matter. A dark patch surrounded by an outer pale ring normally reveals such disused pits. Another consideration is where a farmer has allowed a manure pile to stand and mature; this can leave a dark patch in the field for many years, due to the leached nutrients.

As always, however, if you have the slightest suspicion about a feature you should in all cases investigate it. The path of an ancient, now-lost river may also appear as a white or dark line in the soil, perhaps associated with shell fragments. Some areas of Cambridgeshire are well known for these features singularly known as a "rodon", as they are for huge pale areas associated with long dried out meres or lakes.

In many areas of Eastern England during the last part of the 19th century there was a widespread industry based on mining coprolites. These are nodules of fossilised dinosaur droppings and other organic matter high in phosphates. Mined in huge open cast pits, coprolites were extracted for use as fertiliser; these pits can still be seen as crop marks and normally slightly darker than the surrounding soils.

Yet another factor to be aware of associated with signs of habitation are small depressions in fields. In many areas of settlement the residents sunk a well for their fresh water supplies. These are now mostly blocked; however, we have seen several that, after rainfall, have collapsed inwards. One example left a crater 20ft wide by some 10ft deep that just appeared overnight. Previously, we had frequently used this as a place to shelter for a cup of tea - so always be cautious of these types of initially small depressions!

Soil Types

Becoming familiar with soil types can also be invaluable. Combine this with your research and you will eventually make knowledgeable decisions about the potential of good search areas. Familiarising yourself with the soil types of your locality should enable you to spot both geological and settlement based variations.

The fertile rich, usually dark valley soils, encourage settlement and have done so for thousands of years. Two principle reasons for this are the high yield in crops and the ease of ploughing. Heavy hill top clays were not usually settled because it is only in the last 200-300 years that machinery has been developed able to cope with ploughing this heavy soil.

However, there are always exceptions to advice and clues and we know of at least three Roman sites situated on heavy clays. This leads us to believe that the farming activities of these sites were conducted some distance away, and to a high degree these were residential and storage areas. In relation to soil types, Ordnance Survey produce a series of maps that show geological distribution over the UK. These are immensely useful; when you have discovered a settlement on a certain soil type you can use these maps to study the extent of that soil variety and may find further settlements. This works particularly well with Roman farmsteads, as these moved around frequently due to the ignorance of soil nutrient exhaustion caused by crop growth.

Crop Markings & Lumps, Bumps & Ridges On Meadowland

Sites of buildings, round barrows, ditches and other soil disturbances can be particularly in evidence through crop markings during long hot dry summers. Varying climatic conditions can make crop marks appear in fields, where there has been no appearance before. A few years ago, on land we were very familiar with, a whole series of Neolithic ringed enclosures just appeared, attracting the local archaeological group to trial trench them.

Aerial photographs are a good way to find similar cereal crop or grassland markings. However, this does necessarily mean that they were visible at the time the photograph was taken.

Sometimes, if you are up on high ground, you can gain a good view across low lying potential fields; using binoculars, as stated before, is highly recommended for this situation.

Dependent on whether there are walls or ditches beneath, in crops such markings will show up lighter or darker with different levels of crop growth. Sometimes you will be lucky and see clear markings. However, if using some of the photographic Web site techniques of looking at areas of known settlements or villas etc, you will be quite surprised at some that show up only as large darker areas of crop growth, with no clearly defined markings at all. This factor alone will assist you in discovering other sites.

There are several reasons why temporary false crop marks may also appear, such as where the farmer has suffered a spillage of grain; this will be defined by a much denser than normal crop growth. Equally, where a spillage of fertiliser has occurred this will have the same effect.

Fig.6.6. A freshly excavated drainage ditch. The dug out soil is ideal for detecting.

With experience though you should be fairly capable of discounting these in terms of size and shape. Remember also that animal pens and enclosures - used up until recently but then demolished - can also result in superb crop/grassland markings that are of little significance to the detectorist. On some large shooting estates the gamekeepers erect huge pheasant breeding pens. These are often relocated, but the scars left on grassland etc where they were positioned can look very much like a large Roman courtyard villa, particularly when viewed on an aerial photograph.

If you are walking in meadowland and notice that the field surface undulates in a series of lines, these are almost certainly strip lynchets. These are caused by ancient ploughing techniques building up soil lines. In this country they most likely to the medieval period, but some Roman examples are known. These may also be known as "ridge and furrow". As the sun sets in the late afternoon some fields can have a striped appearance, where the linear depressions fall into shadow. These lynchets are evidence of ancient agricultural work, and very much a sign that there will be some degree of nearby settlement.

Often in association with this type of earthwork you will notice other varieties of lumps, bumps, and ridges. Some of these may be drovers' ways where herdsmen guided their livestock to market. Others may be hollow ways, which are old tracks that can have very ancient origins. Many hollow ways and tracks can be evident in areas where a later village became detached from the church. Another type of settlement indicator to be aware of is the "baulk" often associated with Iron Age settlements. In many cases this is noticeably evident where a field edge drops sharply several feet for quite a distance. The baulk can be semi-circular, nearly completely circular, or simply a fairly straight linear depression. When observing such features, often something will nag at you such as, "It's not geological in this area, it must be man-made." Often these baulks are used as modern agricultural tracks.

Sites where ancient windmills once stood are still often emphasised with a slight mound, sometimes referred to as a "tump". It is worth remembering that sometimes ancient windmills were established on an already existing feature such as a Roman or even earlier burial mound. We found a superb example of a "tump" in meadowland that was ploughed for the first time since the Second World War. Unfortunately, the top of the mound was thickly covered in flints, pottery and tile. It was impossible to detect on the immediate site due to a dense covering of collapsed building debris, but around this finds were very prolific.

Chalk Pits & Chalk Quarries

In the past, particularly in Victorian times, the digging of chalk pits revealed many ancient habitation sites and cemeteries. Some chalk pits may even have Neolithic flint mining origins. Perhaps of more interest to the detectorist is that the Romans also excavated many such pits. This is evidenced in some areas by open pits and filled depressions alongside Roman roads; these pits once provided quarried flint for road metalling. Today, many of these pits are simply overgrown or ploughed out hollows. However, there are many pits that are still open and provide a deep cut face into the local soils.

One such pit that we investigated has a crumbling chalk face. Recently, a huge portion of this collapsed revealing a cross section through an Iron Age refuse pit. This was a real surprise as nothing from this period had been located here before. The pit itself was about 4ft deep and consisted of layers of burnt cobblestones, broken pottery, and pig and

Fig.6.7. Ernie searching a medieval site on which the house platforms and ditches have been flooded by heavy rain.

goat bones. Some pottery still showed scorch marks from cooking and was in perfect condition. From among all this refuse we found a perfect handmade clay loom weight, still showing finger print markings from the person who had moulded it over 20 centuries before. Our next task is to locate the settlement associated with this pit.

Please use extreme care and caution when visiting such sites, and preferably go with a colleague. Exposed vertical chalk faces have a tendency to be very unstable. When quarry work is in its early stages the topsoil is often removed from huge areas. In localities where the topsoil is only a few inches in depth this means that foundations and ditches etc would actually be cut into the chalk. When these are neglected and are finally demolished or the top structures eroded away, the remaining features will often re-fill up with topsoil.

The topsoil filled sunken sections of Anglo-Saxon "Grubenhaus" style structures particularly exemplify this, as they can appear as clusters of oblong darker patches. Other features that can regularly show up in this variety of topsoil removal include postholes of timber framed buildings, and cremation cemeteries.

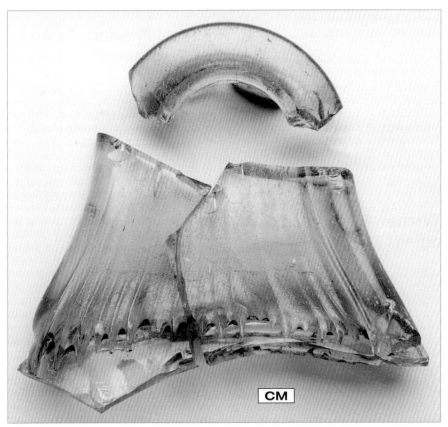

Fig.6.8. Fragments of a Roman glass vessel found on a Roman villa site.

I can remember visiting the site of such a cemetery that dated to the late Iron Age. The topsoil had been skimmed away leaving hundreds of small round dark patches at about 8in depth. Upon closer examination, each darker area consisted of burned bones, and scorched flints with traces of charcoal. No pottery fragments or even any associated artefacts were found on this site, so it appeared that the bodies had simply been cremated and their ashes collected and deposited into holes in the ground.

Therefore on a white, chalky or pale background you can on occasion clearly see outlines of buildings, ditches, barrows etc. Although we are primarily concerned with chalk here, any area subject to extreme soil removal (eg sand or gravel quarries) are all equally well worth investigating.

New Road Routes Involving Cuttings, Pipelines & Building Sites

Seeking permission to search these types of locations for ancient features is also very worthwhile; as always be on the lookout for anything slightly different in colour, texture etc. As already mentioned, local newspapers are the best source for keeping track of new developments.

If you locate anything that you believe may be of serious historical interest you should immediately report your findings to your local museum or archaeological unit.

Any old or ancient ditch, foundations etc that intrude into the soils, clays or chalk will almost always fill up with new topsoil as they become deserted and derelict.

With regards to particularly ancient flint mines etc, later glaciation could be responsible for filling them with clays and pebbles. Concerning cuttings and new verges these can often truncate signs of ancient activity. Different deposits of soils and clays or stones may be able to help you date certain features.

Always keep an eye open for other signs of activity (ie tiles and oysters that may appear in the band of top soils at the uppermost section of the cutting). Also look out for areas of burned earth; if quite large these could indicate the presence of kilns. If they are smaller burned patches and widespread, they could indicate the sites of hearths or even a cemetery with, therefore, hopefully associated settlement.

Other areas worthy of searching along are new pipelines. These initially involve an often-deep trench being excavated to lay the pipeline. Searching and checking the spill from these excavations can be very productive. Such a pipeline in our locality revealed the foundations of a Saxon hut as well as two Roman inhumation burials. Several basic earthenware pots accompanied the burials, as well as a wonderfully decorated samian bowl.

Drainage Ditches, River, Pond & Moat Dredgings

In areas that are prone to flooding, many farmers dredge out frequently or even excavate new ditches alongside field edges. The removed soil can be well worth examining. One such ditch we examined contained some superb examples of Roman and medieval pottery fragments. These included colossal storage jars and *amphorae* handles many still with the potter's complete fingerprints remaining on them.

Look out for rivers that pass through or near sites of interest. These are sometimes dredged, and numerous interesting items are often brought to the surface by this process. Such dredgings are often spread near the river but in some cases can be removed and deposited a considerable distance away.

The same principle applies to moats excavated in medieval times. Adjacent to these, and the enclosures they delineate, the soils can be totally different to those normally found in that area. This is due to the deposition of excavated soils, clays, gravels etc. Such moats that have been dredged since construction can yield some excellent metallic finds as well as un-abraded ceramics.

Many village ponds have also been dredged in recent years. The dredgings are usually dumped locally, more often than not in the corner of an agreeable farmer's field.

One such local example, when newly deposited, yielded over 20 superb condition copper kettles, and still to this day releases good examples of white metal Victorian commemorative medallions.

Over the years we have read, seen and heard of complete Roman pots, beautiful un-patinated bronze coins, and Saxon as well as Viking artefacts coming from ditch/river based soil deposits. Such areas are always worth keeping an eye out for, as the artefacts they bring to the surface are often in wonderful condition, not having been subject to any agricultural disturbance.

Animal Presence As An Indication Of Past Human Activity

Animal presence can be a very good way of helping to assess the potential of a site. Rabbits and badgers often dig down to great depths; never miss an opportunity therefore to search the soil spill near their holes.

So far we have seen tiles, oysters, and pottery fragments evident in such soil spills. There is one thing we have noticed that we believe might be of interest to fellow detectorists. On numerous Roman and medieval sites, after ploughing and rolling, there is a great deal of mole activity. Flocks of crows, gulls and pigeons also congregate on such sites. The only reason for this seems to be that past habitation has changed the soil type into a richer organic medium. This is most likely caused by many, or even hundreds of years, of deposition of waste organic matter. This, in turn, has led to a greater amount of earthworm activity than in the surrounding, often-heavier soils. The moles and birds are obviously taking advantage of the increased food source.

Molehills are also worth checking as on habitation sites they are often packed with pottery, tile and decayed plaster fragments. On some Roman and medieval sites where there is extensive subterranean decayed plaster deposits, mole activity can assist in creating a much paler appearing area in the soil. This is due to plaster being brought to the surface, and in some cases further distributed by agricultural activity.

While you are out in the countryside, the calls and presence of ducks and moorhens can lead you to discover small hidden stretches of water. Some will be ditches or natural ponds, but others may be moated sites or defensive earthworks worthy of further research.

Some years ago, the spotting of several moorhens feeding alongside a wood led to us discovering a tiny moated site. Amazingly, this did not feature on any maps or records we had looked at. Therefore when I look at the marvellous finds made in the area - including four lovely hammered groats - it is to the waterfowl that I give thanks.

The Presence Of Plants & Trees

Plants and trees can often reveal soil type, and therefore can be good indicators of settlement. Recent house sites and Victorian dumps can abound in ground elder. On chalk downland the common nettle can be an indicator of activity. We know of one ancient house and several old farm sites on grassland

Fig.6.9. Some of the pottery and building debris found on a villa site.

that can almost be mapped out room by room by their associated nettle growth. Willow trees and alders indicate areas of wetland that may have been ancient ditches or moats. Sometimes ancient gnarled willow trees can still show the course of a river that has long since dried up; searching along these could reveal many finds.

The shape of some trees, such as that caused by pollarding is also indicative of human activity. Some ancient hornbeam woods, as well as willow trees, were pollarded hundreds of years ago. With such a strenuous, labour-intensive activity many losses would have occurred.

Pollarding is noticeable by a thick tree trunk, often topped in a swelling crowned with many off-shoots. If your interest is in items of Georgian, Victorian or Edwardian finds look for traces of old orchards. Pears and some *prunus* (plum) varieties spread well and live to a great age. Even sparse evidence of such cultivated fruits may indicate the garden of a long demolished house. Many years ago the finding of a Victoria plum tree, in the middle of nowhere, made me look further. I then noticed some ancient pear trees, and a solitary walnut tree. Looking around the site, I spotted some old bricks and willow pattern china fragments below the nettle growth and even an old tin bath in the hedge. Searching around the field edges abutting this area uncovered some of the best ever Victorian and Edwardian coinage we have found. Sometimes it

takes just a single slightly unusual feature to make you stop and examine the potential of the area upon which you stand.

Field Surface Flooding

On both meadowland and ploughed land, we have noticed that areas where collections of individual small houses or a villa etc have stood tend to be marginally shallower than the surrounding levels. Sometimes these shallow basins are only noticeable if you crouch down and look sideways along the field contours. It seems that after a while animal and human activity compacts the floors and surrounding paddocks etc.

Many of these sites are obvious in wintertime where, with severe rainfall, shallow flood pools are often to be found exactly where the buildings stood. Of course, after flooding other factors become enhanced such as pottery, tile and stone scatters. Other features prone to surface flooding which reveals their presence are old boundary/defensive ditches, Roman canals, strip lynchets, ploughed out tracks, and the sites of ancient filled in ponds. Mediaeval moated sites mainly exist these days hidden away in small areas of woodland and meadow. Even though they can be dry most of the year, a few hours of rainfall can refill them again, making them temporarily more obvious - particularly when partially obscured by summer undergrowth.

Worked Flints, Glass & Bones

Almost every landowner or seasoned detectorist I know has found at least one example of a worked flint. Fieldwalking will obviously allow a good chance to experience and collect a huge variety of non-metallic items. Not long ago I saw the most beautiful Palaeolithic hand axe simply picked up by a farmer as he walked over his land after a recent rain shower. This same farmer also has one of the best private collections of fossils that I know of ranging from dinosaur vertebrae to a section of mammoth tusk.

Glass is another substance that is often encountered. Whether it derives from a Victorian bottle dump, is a fragment of Saxon bead, or part of a Roman window, glass is a good indicator of activity. With glass there is no real rule of age by colour; more often it is texture and bubble formation that helps to date it.

It's quite amazing on old dumpsites and edges of large houses how many relatively recent Sunday joint bones one can encounter. Meat consumption for humans goes back probably well over a million years, and therefore butchered bones are a vital clue in the search for habitation. Bones - unlike the sun bleached white oyster shells - are prone to colouration and chemical change influenced by the soil type

they are deposited in. Bones from acidic peat soils can be brown to almost black, and on occasion can be very hard to date scientifically due to mineral absorption. Examples found on agricultural soils are slightly easier to date. In our experience we have found that bones relating to medieval and Roman sites are mostly creamy white and have a hard calcite texture. The human bones I found in association with an Iron Age cremation actually "chink" when moved as if made from stone. Iron Age sites that we detect upon have widespread scatters of pig and goat bones, while most Roman sites we know of seem to be strewn with horse bones, and strangely these are mainly sections of broken rib. One such bone found still bore the knife marks from when the animal was butchered.

Fieldwalking & Crashed Wartime Aircraft

It may be that your particular interest is finding relics and artefacts in association with crashed wartime aircraft. This still represents a large section of metal detecting activity in the UK. In searching for the whereabouts of such an incident, many previously mentioned factors and information sources can

Fig.6.11. An overgrown bomb crater from an American 100lb bomb.

be used, but there are also a number of different issues to consider as will be outlined briefly.

Most wartime crashes were photographed and there are a number of sources where you can still obtain copies of the prints. The presence of a photograph can be invaluable in pinpointing exactly where an aircraft came down. With all historical incidents traced by using contemporary photographs, don't forget to take your own modern day shots of the same scene if possible. Comparing photographs to see how much or how little a scene has changed can be fascinating, and will be of great interest to future researchers.

I know of many detectorists who have actually discovered the site of such a crash through searching an area for un-related artefacts such as Roman coins. If you are seriously thinking about taking up this side of metal detecting as your main interest, there are several things to consider. Firstly, all wartime relics associated with crashed aircraft technically belong to the Crown (even German ones) pursuant to the Protection of Military Remains Act 1986.

No one minds you finding a few scraps of alloy etc; however, if you are considering a serious excavation based upon your metal detector finds you will need to apply for a license from the Ministry of Defence. All such buried remains are subjected to the Protection of Military Remains Act; therefore this license must be obtained before any excavations are progressed.

Obviously, the first consideration in location is to attempt to see if any local residents remember the incident. If they do, they may know precisely where your aircraft crashed, or at least an approximate area. Even an approximation places you a great deal further ahead in your research than previously. You may be surprised at how many artefacts were "liberated" from crash sites that remain tucked away in drawers or barns even today. So get yourself known by local people and if you are successful you will begin to be shown many such items.

Woodland Crash Sites

Due to the inaccessibility of most types of woods during the Second World War (and ever since), it is still possible to find quite large pieces of aircraft in them. However, with increased interest in such relics the amount of these pieces available diminishes with each passing year. If you know that the particular aircraft you are interested in crashed in a wood, several factors need to be taken into consideration. For example, are there any small to medium sized areas of different plant or tree growth? Often a burning aircraft will seriously affect the stability of soil. Due to the presence of oils and other pollutants it might take years for growth to recover.

In mature woodlands observe any large gaps in the canopy. In and around these areas try to examine

Fig.6.12. *General aircraft crash site debris found by fieldwalking.*

the ground for fragments of metal, rubber etc. Patches of blue crystalline powder are a good indication that you are in the right area, as this is probably aluminium oxide. This results from aluminium that has been weathered or buried for quite a time. Also examine tree trunks and large branches for any obvious structural damage (however, remember that the 1987 storm/hurricane damage is still very much in evidence, so be careful in your conclusions). In some cases severe tree damage may actually be in association with fragments of aircraft metal actually protruding from these areas. I once saw a section of propeller blade projecting from the trunk of a mature beech tree, and at the time of observation the crash had happened 52 years before. If you are lucky, therefore, and sharp-eyed you may still find such graphic evidence.

It can also be very revealing to pass the search head of a detector up and down tree trunks in both actual and suspected crash site areas, but do not damage living trees. While the bark of most coniferous trees heals well from small wounds, the inner wood does not. If there are any dead conifers in the area you are investigating try peeling away some bark; the tiniest abrasion caused by impacting metal etc will show up on the inner wood surface really well.

Metal detecting on a local B17 crash site within woodland revealed hundreds of fragments of airframe, along with crumpled cockpit instruments. Of all the numerous finds the most poignant was a twisted and crushed Lieutenant's cap badge. This was later shown to an eyewitness, who found it astounding that 59 years later he was able to handle such an emotive artefact from an event he had seen as an 11-year-old boy.

Crash Sites On Arable Land

If when wandering a field you come across a profusion of twisted aluminium, once molten metal globules, Perspex, rubber, bullet cases etc, the chances are that you have discovered the site of an aircraft crash. Dating the crash can be reasonably well ascertained from the items that you may find. Bullet cases, for example, are quite often dated although you need to be aware the bullets may have been in storage for some time before use. Sometimes serial numbers can be found on maker's and manufacturer's plates. Small stamps applied to spars and other parts of the airframe etc can often tell you who the manufacturer was, and may also bear dates.

Similar factors discussed previously also apply to air crashes such as differences in soil coloration, stone scatters, and perhaps crop marks. Where the soil has been burned this may show up as a darker patch. I know of a local Lancaster crash where you can still see the difference in soil where the four engines impacted the ground. If the aircraft penetrated the surface and either it or on board bombs/fuel tanks exploded, this will bring up different strata of stones or soils. These can remain visible for many years, and may encourage a different level of crop growth or weed infestation. Even today on some crash sites, particularly in heavy clay districts, it is possible to take a pinch of soil and still smell grease and aviation spirit in it.

In some ploughed fields with chalky soils, grease from aero-engines etc degrades into a bright purple or pink substance, which is highly visible. Check any ditches alongside known crash sites, as in the last 60 years many large parts of aircraft have been ploughed up and removed to adjacent ditches. Numerous finds are made in such places. I was recently investigating the crash site of a B24 Liberator and in a nearby ditch spotted 5ft of stainless steel belt linkage once connected to one of the 0.50in calibre machine guns. This aircraft had exploded with such violence it had blown away a huge section of hawthorn hedge; this has never grown back.

In the course of detecting anywhere always be cautious of large unidentified buried or partially buried metal objects that you uncover as many wartime bombs, grenades and rockets remain undiscovered. The larger bombs are either accidentally dropped Allied bombs, deliberately dropped enemy ones, or those from both sides still trapped in underground lying aircraft wreckage. Obviously, the chance of this type of find increases around Kent, outer London and other heavily bombed areas. Should you find any unexploded ordnance, mark the spot and then inform the landowner and the police as soon as possible.

Always treat anything suspicious with the respect it deserves, as some old explosives deteriorate and become highly unstable. In reporting the find you have carried out an act beneficial to the public, as the dangerous item will be disposed of.

Crash sites themselves should always be treated with the greatest of respect, particularly where

Fig.6.13. An SC50 German bomb from a Dornier 17 crash site - found by a metal detectorist.

human lives have been lost. Should you find personal items or effects, such as human bones or rings, wrist watches, identity tags etc you should notify the police and the Ministry of Defence as soon as possible. Even after 60 years such finds are still possible. I recently saw the wallet from a P47 Pilot killed in 1943. This came from a licensed excavation, based on a previous metal detector assessment. Remarkably, when opened there were still photographs of his girlfriend and two penny postage stamps inside, as well as his car keys.

Although this is only a small section on aircraft crashes we hope that it has maybe broadened your horizons into yet another aspect of the fascinating hobby of metal detecting.

SEARCH TECHNIQUES & METHODS

Nearly all types of modern metal detectors are ergonomically designed for balance and comfort in order to minimize arm strain. How you use your machine is therefore down to whichever technique you feel most comfortable with. Some people swing their detectors in a wide sweeping arc, while others simply sweep from side to side in straight lines as they move forward. But, whatever style you adopt, the most important thing to remember is that you must keep the search coil as close to the ground as possible at all times. Never swing the detector as if it were a pendulum, as this will limit the detector's depth-seeking capability to the centre point where the search coil comes closest to the ground only, rather than across the whole sweep of the arc.

It is also important that you sweep the search coil slowly; going too fast will dramatically reduce your find-rate.

Search Techniques

If you are searching a field for the first time and want to assess it as a potentially good or poor site, or have only a limited amount of time available due to seeding about to take place, it might be a good idea to adopt an explorative search technique.

One effective method used by many detectorists is known colloquially as the "Union Jack" system. This involves simply detecting across a field from corner to corner both ways to create an "X" and then searching around the outside perimeter. The idea behind this is to quickly ascertain whether the field contains any "hotspots" or productive areas that might be worth concentrating on. This is particularly useful in the case of some Roman sites, which can be very localised and contain finds limited to a small area.

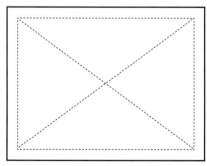

Fig.7.1. "Union Jack" search system.

If you do find a small but productive area, you can then adopt a more methodical search pattern in order to maximise your find rate. This can range from the extreme of pegging out the ground with lines and pins, to simply using some large stones or dead branches as markers. Intensive searching involves slowly sweeping the ground in small straight overlapping lines and then covering the same area in the same way from a 90 degree angle (Fig.7.2.). This technique is known in the hobby as "criss-crossing". It is very effective when a scattered hoard has been located and the plough has spread the coins over a distance from their original burial spot.

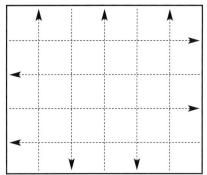

Fig.7.2.

Digging

When a target has been located and pinpointed by the search coil, take great care not to thrust your spade or trowel into the ground at the spot indicated. Start digging about 6in away from where the detector has pinpointed a target and dig towards the point of maximum signal strength. Many detectorists have scratched and effectively ruined a good coin or artefact with the blade of their trowel by not following this simple rule.

Another point to remember when digging is that you should always fill the holes in afterwards, whether you are digging on anything from a grassed common to a roughly ploughed field. Many landowners make this a condition when granting permission to detectorists to search their land, but it is also simply good manners. On pastureland, holes left unfilled can be dangerous to livestock, as well as annoying for the landowner! Always make a point of leaving your site in the condition that you found it.

Soil Conditions

Soil conditions play a major part in achieving good results with a metal detector. To gain optimum performance from your machine the ideal conditions are flat compacted soil, which is also reasonably moist. This is often the case with pastureland but conditions on plough soil will vary according to the time of year.

Not many detectorists will venture onto land that has just been freshly ploughed as the deep furrows make detecting extremely difficult as well as physically uncomfortable. Try waiting until the farmer has levelled off the soil with a harrow. If you are lucky, the farmer may even allow you to continue detecting after the field has been drilled when conditions for detecting are even better still.

Fig.7.3. Roman silver denarius of the emperor Caraca II found at a depth of over 9in on moist, well-compacted soil.

Moisture in the soil is important, as this will improve the conductivity of buried metal objects and the depth at which they can be located. To see this for yourself, try detecting on soil in very dry conditions and then see the difference when you return to the site after plenty of rain has fallen. Many detectorists swear that their depleted sites came "back to life" after weeks of rain.

Mineralisation & Discrimination

Mineralisation can be caused either by natural elements in the soil or by contamination due to human habitation. For example, many large Roman sites had iron workings, which serviced the needs of troops and their horses etc. As a result, the sites tend to be littered with iron scraps of all descriptions, particularly nails and dross from the furnaces. This can cause annoyance as well as a lot of unnecessary digging.

Most modern metal detectors are equipped to counter this problem by means of a "discrimination"

facility. If you encounter iron contamination on any site then try adjusting the discrimination by turning it up by a few increments until you reach a point where the iron is rejected. Don't be tempted to use full discrimination as this will force the machine to reject good finds that have low conductivity levels such as gold rings, hammered coins etc.

If mineralisation is encountered, better depth is often gained by turning down the detector's sensitivity level, even to as low as the half-way mark.

EXAMPLES OF DETECTOR FOUND ARTEFACTS

Whether you are a newcomer or a seasoned veteran to the hobby of metal detecting, the potential is there to make finds from all ages. One moment you can be admiring the beautiful green patina on a Hadrian *sestertius*, and minutes later puzzling just how a 1971 new penny can be in such appalling condition. Or, just having recovered a hammered silver penny of Edward I, right next to it you might find a spent .303 cartridge from the Second World War.

Metallic losses have occurred in the UK for nearly 4,000 years, and this explains the vast time span in finds that you can encounter. Successful metal detecting is a combination of research, technical equipment and, of course, luck. Luck is one of the biggest players in the search for artefacts as it respects no boundaries and has no rules of logic. Lucky is the young boy who having purchased a budget model detector finds a hoard of Roman silver *denarii* on his very first venture out (it has happened) and lucky is the finder of a superb hammered gold coin found on a site that many detectorists had said was "thoroughly played out".

On this subject "lucky" is also the detectorist who realises that among all the rubbish he will inevitably collect, the statement "worked out" is truly worth discarding. This was exemplified recently in a field, which is fairly local to us. This field has been detected upon probably for over 20 years, and yet a hoard of silver and gold Roman items were discovered there. Incredibly they represented votive offerings to a new unrecorded goddess called "Senua". What a marvellous contribution to our history!

Lucky also was the middle-aged lady who after 10 years of detecting found a spectacular Bronze Age mirror on a field she had been meaning to look at for ages.

Sometimes you may experience a "run of luck" where, to everyone's amazement, you simply cannot stop making finds. I have experienced one of these where I unearthed a Saxon *sceatta* and a Celtic silver unit within 10ft of each other, combined with over 50 other coins (all found in an hour).

Many clubs or groups also seem to have one individual who seems to find more than everyone else even if they are using the same type of detector; this can only be attributable to luck.

Research and equipment are essential, but spending vast amounts of time and money do not guarantee good finds; it takes luck to guide your search head over the target.

The finding of a superb hoard may be missed by only a few inches. That rainy day when you ponder whether to go out or not, might make a difference

Some of the detector and "eyes only" finds on display at a Federation meeting of metal detector clubs.

between finding out who has done what in the latest soap opera or finding a handful of shining gold staters.

It is this element of chance that keeps most people detecting. Although the majority of us may never find a Hoxne or Shillington hoard, only by detecting do you increase your chances of doing so.

Although not on the same scale, many of us find treasures every time we venture out. These may range from a rare military button to that sought after trade token you have always wanted.

The Pastfinders have, to date, never found anything comparable to some of the major UK discoveries, but we have had some superb important individual finds. Several years ago Tony Roche, a Pastfinders member, found a unique coin of the emperor Gallienus, and this coin now forms part of the British Museum's collection.

We were also lucky enough to find the fourth known example of a Roman cube matrix, our

example showing the portrait of the Roman Emperor Philip I.

The following is a selection of "what's out there in the fields". The concept of this chapter is not to provide huge sections of informative text, but to give more of a visual appreciation of related finds. We hope that in addition to possibly being used as reference for your future finds, this section will stimulate people to consider, take up and continue metal detecting. The majority of finds illustrated here have been made by The Pastfinders; where made by a colleague or friend full credit will be given.

It is all too easy to take things for granted. However, there will never be any more Georgian coppers, Roman brooches, or hammered coins manufactured. What lies in our soil now is our heritage and must not be allowed to decay due to others' ignorance or bias. We hope that the importance of metal detection will continue to be granted the recognition that it deserves: the recovery of the past, conducted today, for the benefit of the future.

Bronze Age Finds

This period of history is known for the manufacture of the UK's most ancient metallic artefacts. Therefore for many detectorists these artefacts can be high on the "want to find" list.

Bronze is an alloy of copper and tin in varying percentages for each alloy mix. Research conducted on some bronze artefacts found in the UK has traced the copper part of the alloy to originate from Switzerland and the tin to Cornwall. This is indicative of well-established trade routes even in those times.

The earliest known bronze axe heads from this period are known as flat axes. These are simple solid cast bronze axe-shaped tools, and are not common finds. Fragments of later hollow axe heads, spear tips and small pieces of copper bun ingots are, however, relatively frequent metal detector discoveries.

Of these, it would seem that the crescent-shaped fractured cutting edge of the hollow cast axe heads is perhaps the most common in many areas. Presumably, the fact that they were cast meant they were quite brittle. It seems logical to suppose that, although superior to flint, many cutting edges broke and became embedded in standing and felled timber. Further research suggests that the cutting edge of a bronze axe would last approximately only 20 minutes when used against most UK wood types.

In many areas fractured bronze spear tips closely follow the frequency of finding these axe head fragments. Presumably these tips, again being quite brittle, would snap when thrown into hard soil, or resistant animal or even human bones.

Another fairly common find are the sections of broken sword blade that have been deliberately snapped. Whether this was for practicality in transportation for future smelting or was a votive act is a matter for debate.

When you do find a complete artefact (or even a fragment) from this period, it seems to hold a distinct fascination. It is incredible to consider that the item may have been lost or buried 1500 years before even the Roman Legions marched along our trackways. This does, however, help to put into context just how old these artefacts are.

The other notable finds associated with the Bronze Age are, of course, hoards. These can be quite small, consisting, perhaps, of just two axe heads and a few lumps of copper bun ingot or similar. However, larger founder's hoards have been discovered containing huge quantities of items including many varieties of axe heads, spears, chisels, gouges and a whole range of accompanying broken items. For example, the recently discovered Bronze Age hoard on Jersey included over 100 individual artefacts.

It is generally accepted that these hoards are collections of worn or broken tools hoarded by a bronze smith to be smelted and re-cast. The many complete and fragmentary raw copper bun ingots in these hoards appear to prove this. One copper bun ingot fragment that we uncovered had been used either as a hammer or perhaps an anvil. This unusual use was indicated by the two flat edges showing numerous indentations, and the sides being split with pressure cracks.

On very rare occasions Bronze Age items manufactured from gold have been recovered. Usually, these are thinly beaten sheet gold dress adornments, solid neck torcs, small pieces of worked items, or raw bullion. It is worth noting that if you are lucky enough to find a hollow cast axe head, on some

Fig.8.1. Bronze Age palstave dating from circa 1500 BC.

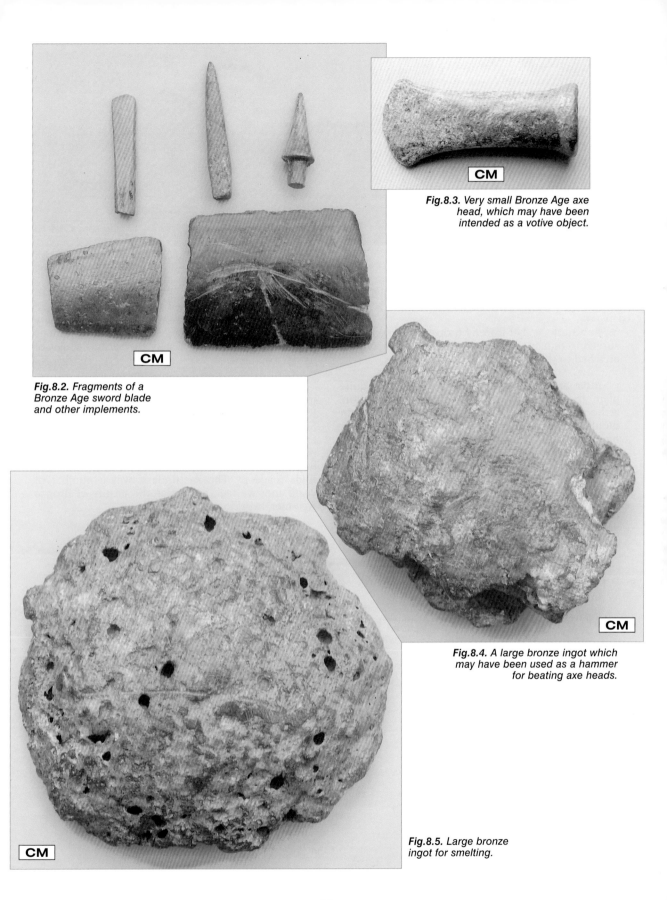

Fig.8.2. Fragments of a
Bronze Age sword blade
and other implements.

Fig.8.3. Very small Bronze Age axe
head, which may have been
intended as a votive object.

Fig.8.4. A large bronze ingot which
may have been used as a hammer
for beating axe heads.

Fig.8.5. Large bronze
ingot for smelting.

Fig.8.6. A single-looped palstave with beaten down flanges.

occasions small lumps of gold bullion have been found in the compacted soil inside them.

In waterlogged or frozen European find spots, the surviving remains of the wooden shaft can still be attached to the axe head.

Most bronze artefacts from this period are quite plain in style with little evidence of decoration, although some axe heads have three or four moulded decorative lines on their sides, or rarely lines and loops.

The development of Bronze Age tools and implements is now very well recorded, and in most cases such finds can be dated reasonably accurately according to style.

The latest development of the Bronze Age axe head is known as the "palstave". On this late variety of axe head the various blade sizes, attachment loops and side wings help to classify the various stages of its development.

This is truly an evocative period of history with its mysterious barrows, ringed enclosures, and the first metal tools and weapons to be used in Britain. Should you be lucky enough to recover a metal find from this "dawn of time" era, then you will probably never forget when and where you found it. The items shown represent a collection of Bronze Age finds made by The Pastfinders over a number of years and, we feel, epitomise the beauty of Bronze Age metalworking.

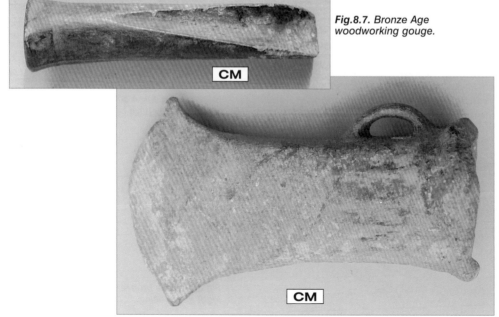

Fig.8.7. Bronze Age woodworking gouge.

Fig.8.8. Socketed axe head.

Celtic Finds

Sometime around the middle of the first millennium BC there was a gradual transition from the Bronze Age into the Iron Age period.

Just as with the Bronze Age, this has been divided by academics into three distinct periods: the Early Iron Age, Middle Iron Age, and Late Iron Age. It was also a time when metal of several different types was used much more frequently than in the earlier Bronze Age.

Although bronze was still used for tools and weapons in the Early Iron Age, the more durable iron eventually replaced it. Bronze, however, still remained one of the principle metals for use in the manufacture of objects such as jewellery, adornments, fittings for belts and harnesses, clothing fasteners, figurines etc.

The Late Iron Age also saw the introduction of the first coinage to be used in Britain. These coins were mostly imported from the Continent and consisted of crude, often concave discs, decorated with such designs as stylised heads, animals, stars, crescents and pellets.

Coin production in Britain probably began somewhere around 40 BC, and just like the Gallo-Belgic issues before them, depicted images of animals (mostly horses), crescents and pellets, etc. These coins or "staters", carried no legends, so are known as "uninscribed staters".

It wasn't until the Romans began dominating Western Europe and bringing their literary influences with them, that names began appearing on British coinage. These were generally the names of tribal kings or chieftains.

Fig.8.9. Examples of Celtic coinage, found on one Iron Age site.

Fig.8.10. A selection of Celtic bronze coins.

Fig.8.11. The remains of a small Iron Age mirror, found by Julian Evan-Hart.

Fig.8.12. Brooches such as this
"horse and rider" type, are
scarce finds from the Celtic period.

Fig.8.13. A Celtic silver
coin of Cunobelin.

Fig.8.14. Iron Age
clothing fasteners, or
toggles, are quite
common finds.

Fig.8.15. A very rare "Upavon"
type Celtic brooch, dating from
around the 4th century BC.

Fig.8.16. A gold stater
from the Climping
hoard.

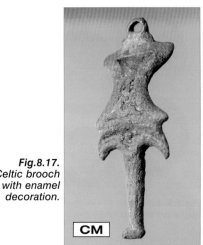

Fig.8.17.
A Celtic brooch
with enamel
decoration.

Although Celtic coins are often found by detectorists, Celtic artefacts are far more rare and elusive. It is probable that many of the latter were broken up for recycling, or buried as offerings to the gods. It could also be that - after the passage of 2,000 years - many artefacts are buried at levels too deep to be detected. In addition, many objects were thrown into rivers, springs and lakes as "votive" offerings, making them difficult or impossible to retrieve.

Many Iron Age settlements became "Romanised" after the Occupation in the mid 1st century AD, and Celtic artefacts and coins are often found on Roman sites. One of the brooches featured, depicting a man riding a horse, is Celtic although it was found on a Roman military site. A simple explanation for this is that many local Celts were enlisted into the Roman army as Auxiliaries. No doubt they were allowed to retain some of the things that represented their own culture.

The circulation of Celtic coinage, however, was banned after the Boudiccan Rebellion, in AD 61, as an act of retribution by the Romans.

An even scarcer find is the brooch featured, which is known as an Upavon type; it dates from around the 4th century BC. This particular example was found on a votive site, near a spring in Hertfordshire. Slightly more common finds from the Iron Age period are clothing fasteners, or toggles. Many of these are decorated, such as the one shown here, although the more desirable fasteners are the enamelled types (also illustrated).

The illustrated Celtic coins and artefacts have all been located with a metal detector from sites in the United Kingdom.

Roman Finds

Although Julius Caesar came to Britain with his expeditionary forces in both 55 BC and 54 BC, it was to be almost another 100 years, under the Emperor Claudius, before the Romans actually returned with the intention of staying. They brought with them a culture that was totally alien to that which had existed here for many hundreds of years.

Along with this new culture the Romans brought new religions, literature, architecture, marvellous feats of engineering and, of course, a whole new range of metal artefacts and coinage. In fact, one could argue that the Roman period offers the widest diversity of metal detecting finds of all the periods of British history prior to modern day.

However, it is not uncommon to hear of seasoned metal detectorists who have yet to find their first Roman coin! This could well be due to the fact that these hobbyists are either "geographically disadvantaged" (many parts of the country weren't occupied as extensively as, for example, East Anglia), or that they haven't done a great deal of research into locating Roman sites.

The most common finds from the Roman period are generally the low-value bronze coins. For most of the occupation these consisted of large coins known as *sestertii*, *dupondii* and *ases*. The 4th and 5th centuries, however, brought many changes to the Roman coinage, which saw the earlier denominations replaced by coins such as the *follis* and the *antoninianus*. These coins themselves were rapidly replaced by even smaller coins, whose names are now unknown but are referred to as: AE1, AE2, AE3 or AE4, in relation to their sizes.

It is ironic that the later much smaller coins tend to be found more frequently than the larger coins, which had been in circulation for much longer. One explanation for this could be down to their sheer size - the loss of a massive *sestertius*, such as that featured in Fig.8.18., for example, would easily have been noticed due to its weight.

Other frequently found Roman coins are the silver pieces known as *denarii*. Owing to their precious metal content, these often survive quite well in most soils. These coins were replaced during the latter

Fig.8.18. Various types of Roman bronze coinage.

Fig.8.21. Roman disc brooch complete with black glass boss.

Fig.8.20. A selection of Roman bow brooches.

Fig.8.19. A Roman "horse and rider" brooch, with much enamel decoration remaining.

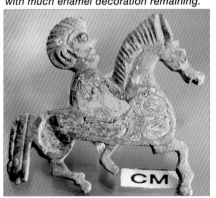

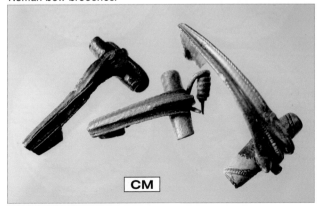

Fig.8.22. A classic example of a Roman knife handle depicting a hound chasing a hare.

Fig.8.24. A Roman "fly" brooch.

Fig.8.23. A unique type of coin, known as a Quadrigatus, of the Emperor Gallienus was found by one of our group and is now in the British Museum.

CM

Fig.8.25. This female figure may have been a Roman furniture fitting.

Fig.8.26. Fairly common examples of Roman silver coinage, known as denarii.

Fig.8.27. A pair of Roman "sandal" brooches.

Fig.8.28. A good example of a late Roman "fish" buckle.

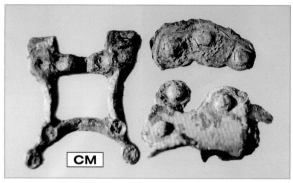
Fig.8.31. Roman military belt and armour fittings.

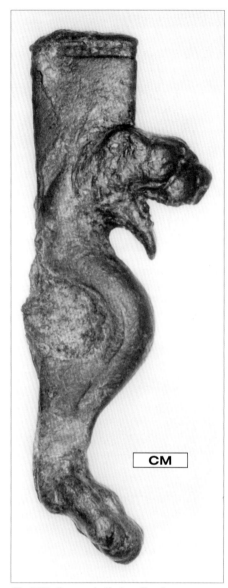
Fig.8.29. Roman knife handle of a scarce type depicting a leopard.

Fig.8.32. A large Roman lead dice.

Fig.8.33. An extremely rare steelyard weight in the form of a "Nubian".

Fig.8.30. A very scarce example of a Roman "swastika" type brooch, found on a Roman military site.

Fig.8.34. Roman keys, such as this, are fairly common finds.

Fig.8.35. A gilded Roman ring with an intaglio, or carved stone.

Fig.8.36. Roman silver denarii.

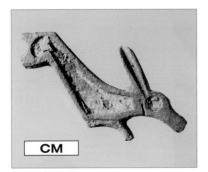

Fig.8.38. A Roman dog brooch.

Fig.8.39. A selection of late Roman bronze coins.

Fig.8.37. Roman umbonate brooch with much of its enamel remaining.

part of the occupation by a much thinner variety known as *siliquae*. Although a much less common find than the *denarius*, *siliquae* also tend not to fare so well on arable soils, owing to their thin flans. If you ever find one without a crack in it, then count yourself lucky!

Perhaps the second most common of Roman finds are the brooches. Again, rarely do these survive intact - particularly the larger varieties - although the range and diversity of types is enormous. The most frequently found type, however, is usually the "bow" varieties, which turn up in considerable numbers on most Roman sites.

The rarer varieties tend to include the zoomorphic (or animal-form) brooches and some of the open work types. Disc-shaped brooches also turn up quite frequently, although many of these tend to be of the smaller, less elaborate, forms.

You may be fortunate enough to find a brooch that retains, some or all of its original enamel decoration. You may be even luckier still, and find an example of one of the more elaborate type disc brooches that were gilded and set with an intaglio or carved gemstone. Roman brooches are often the most impressive elements of a detectorist's collection.

The Roman coins and artefacts illustrated have all been located with a metal detector on sites in the United Kingdom.

Saxon Finds

It is often argued that the Saxon period commenced at the end of the Roman Occupation. As the legions returned to defend their homeland from the Vandals and the Visigoths, Britain lay open to invasion from the Saxon hordes across the North Sea. Also known as the "Dark Ages", this period is the least understood of this nation's history.

Unlike the Romans, the Anglo-Saxons left little for us to study in the way of structural remains. They tended to shun the idea of living in solidly built houses and preferred to dwell in wattle-and-daub type constructions, which apart from a few post holes have not survived the passage of time.

Much of what we know about the Saxons has been from studies of their graves. The Saxons tended to bury their dead in large cemeteries, several of which have been found in recent years by metal detectorists.

One of the few surviving legacies of the period are the magnificent churches, which the Anglo-Saxons built in large numbers across the country. Many of these were constructed using materials looted from earlier Roman constructions, such as villas, the evidence can often be seen as Roman bricks or tiles in the walls of the churches.

The Saxon period provides us with some of the most beautiful metal artefacts that can be found. These are undoubtedly epitomised by the exquisite workmanship seen on the gold and enamel inlaid items, such as brooches, religious crosses, and the strange *aestels*.

Examples of such workmanship can be seen on items excavated from the Sutton Hoo burial mound in 1939. Superb disc brooches are found with chip carving and they are often thickly gilded. Other brooches sometimes found are the square headed type. These can be very large and packed with decoration and panels of inlaid garnet.

Fig.8.42. An extremely rare Anglo-Saxon mount or fitting, depicting a "horned" figure.

Fig.8.40. Very scarce silver strap end.

Fig.8.41. Late Saxon strap end.

Fig.8.43. Possible Saxon pottery decorating tool.

Fig.8.44. Anglo Saxon primary sceatta.

Enlarged

Enlarged

Fig.8.45. A cut halfpenny of William I (AD 1066-1087) of the class "profile right type". Found by Jeff.

Fig.8.46. A fine example of a Frisian Saxon silver sceat. Date AD 695-740.

The most common brooch recovered from this period of history is probably the crossbow brooch. These can be of simple design and most likely will be made from bronze, although silver and gold examples are known. All too often it is only fragmentary remains of these brooches that are found; nevertheless anything Saxon is always a welcome find.

Some very late Roman sites may yield evidence of early Saxon occupation. One indication of this are pierced Roman coins, used by the Saxons to hang around the neck as pendants or to form sections of necklaces. Late 4th century coins are usually used, but we have been fortunate enough to find pierced *sestertii*, *dupondii*, and *antoninianii*.

The typical entwined abstract Saxon artwork is probably best defined on some of the strap ends recovered. These are normally made of bronze with silver niello inlay; however, if you are really fortunate, you may encounter one made of silver.

Some sites also seem to show evidence of Saxon attempts to copy Roman coinage, perhaps as a desperate measure to replace dwindling Roman supplies. Such evidence consists of lead copies as well as cast copies of Roman coins.

The earliest recorded Saxon coins are called *thrymsas* and *tremisses* and are very small gold coins usually struck on a thick flan. These tiny gold coins later on became very debased, and seem to have been replaced by the silver coins known as *sceattas*. Later Saxon silver pennies are larger but thinner, being similar to medieval hammered pennies in size.

For most metal detectorists Saxon finds are scarce, unless of course he/she is lucky enough to locate a Saxon cemetery. The finds illustrated in this section probably represent the most frequent types of Saxon find made by UK-based metal detectorists. The Saxon coins and artefacts illustrated have all been located with a metal detector from sites in the United Kingdom.

Medieval Finds

After the Saxon period we see the emergence of the Normans. Like the Saxon era, very few finds - other than coins - tend to be found from this period.

Many of the medieval finds made by detectorists tend to date from the 13th, 14th and 15th centuries. These can include coins, dress fasteners, belt and harness fittings, seal matrices (in various forms), keys, and jewellery. The coins tend to be the most common finds from the medieval period. These are usually in the form of the small hammered silver pennies or groats. It is not unusual to find some of the pennies cut into halves or even quarters, as this was the most convenient way of providing small change.

Dress fasteners, often in the form of ring brooches, turn up frequently - probably lost by farm-workers toiling in the fields. These come in various sizes from about 10mm across to larger, and more elaborate decorated types with inscriptions.

Precious metal finds from the medieval period are very scarce, owing to the fact that ordinary people were forbidden to wear items made from gold or silver. This privilege was reserved mainly for people of high status. Much of the jewellery and other accoutrements of everyday life tend to be made out of bronze, or even pewter. Copper alloy fasteners and buckles tend to be the most frequently found artefacts from the medieval period. The medieval coins and artefacts illustrated have all been located with a metal detector on sites in the United Kingdom.

Fig.8.47. Medieval buckles.

Fig.8.48. Medieval hammered coins.

Fig.8.49. Medieval seal matrices in bronze and lead.

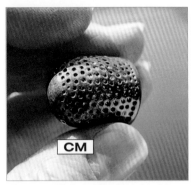

Fig.8.50. Medieval beehive thimble.

Fig.8.51. Medieval strap end.

Fig.8.52. Cut half long cross penny of Henry III.

Fig.8.53. Gold quarter noble of Edward III.

CM

Fig.8.54. Medival ring brooch complete with inscription.

Fig.8.55. Gold half noble of Edward III.

Fig.8.56. This heraldic pendant would have hung from a medieval knight's horse harness.

Fig.8.57. Medieval buckles.

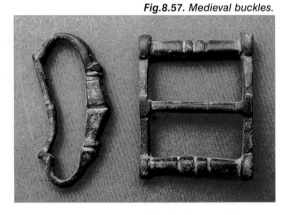

Tudor To Modern Finds

This section covers metal detector located items from approximately 400 years of the United Kingdom's history. Consequently the variety of finds is enormous (not counting the imported Continental artefacts that are sometimes encountered).

Standard types of finds made from this period differ little from other periods, in that you will find coins, fittings, and jewellery. However, this era has an additional large variety of "extras" that really come to the forefront such as: coin weights, jettons, trade tokens, buttons, and the first real numbers of milled coins. In fact, in terms of coinage these 400 years start with the presence of the still often poorly struck hammered coins and finish with the later refined stages of mass milled coin production.

One of the most important events in the period in question was without doubt the English Civil War, which resulted in some of the largest coin hoard depositions ever made. Needless to say, metal detecting has been at the cutting edge of their location. The reign of Charles I saw one of the largest strikings of coin varieties of any monarch. Some of the large shillings, gold unites, and the huge silver pound are among some of the most collectable and sought-after coins as detecting finds. Other very interesting items such as trade tokens and jettons offer a huge range for the collector, due to their sheer numbers and variety.

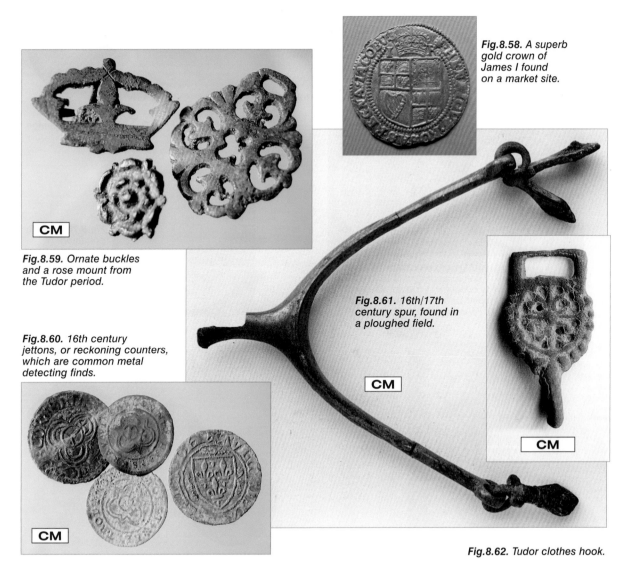

Fig.8.58. A superb gold crown of James I found on a market site.

Fig.8.59. Ornate buckles and a rose mount from the Tudor period.

Fig.8.60. 16th century jettons, or reckoning counters, which are common metal detecting finds.

Fig.8.61. 16th/17th century spur, found in a ploughed field.

Fig.8.62. Tudor clothes hook.

Fig.8.64. Tudor snake buckle with much gilding remaining.

Fig.8.65. Silver thimble, possibly dating 17th-18th century.

Fig.8.63. Medieval quillon dagger chape.

Fig.8.66. 17th century Trade tokens.

Fig.8.67. Early Georgian copper coinage.

Fig.8.68. Due to a shortage of precious metal in the 18th century, foreign coins, such as this Spanish silver coin, were allowed to circulate in Britain. They were usually counter-stamped with the monarch's bust before being issued. Found by Nick from Gosport.

Fig.8.70. A lovely condition George III silver bank token for three shillings from 1811.

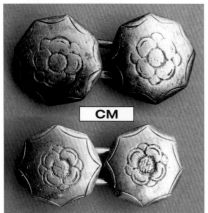

Fig.8.69. Silver cufflinks.

Fig.8.71. Commemorative medallions from the Victorian period.

Fig.8.72.
A selection
of silver coins
dating from the
early 19th to early
20th century.

Fig.8.73. Enamelled badge from the engine
of a German ME110c, which crashed at
Hobb's Cross, Harlow, during 1940.

Fig.8.74. Queen Victoria Jubilee shilling of 1887.

Fig.8.75. 1940s-1950s style model Meccano car.

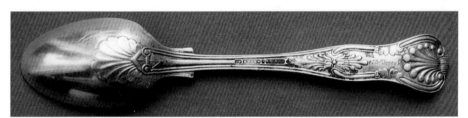

Fig.8.76. Silver plated spoon, found in meadow.

In the 18th and 19th centuries, commemorative medals and tokens abound, and were struck for battles, engineering feats and exhibitions. In the early 19th century there was a shortage of silver for minting coins. The Bank of England silver token issues, as well as countermarked dollars and Spanish *real* pieces evidence this.

It is perhaps not surprising that this period saw a high level of forgery both of these tokens and the short supply standard silver coinage.

The Victorian era is highlighted by a superb series of coins, gaming tokens, military buttons and badges, as well as ornate examples of jewellery.

Leading into the 20th century finds are much the same; however, as the century progresses a greater level of militaria is evident. Almost every field seems to contain military buttons, some being quite remarkable and ornate in design.

The Second World War, due to its nature, placed even more metallic finds in our fields including buttons, cap badges, fragments from crashed aircraft, and spent bullet and shell cases.

One unusual find that I made some years ago when investigating a crashed Lancaster bomber was of a worn and corroded Charles I shilling incredibly jammed inside a corroding fragment of the aircraft's fuselage. I still have the shilling, although the corroding aircraft fragment was too far deteriorated to save. However, the find is the perfect example of the varied and often unexpected finds metal detecting will bring you. Quite how the shilling came to be in this final state is anyone's guess.

The coins and artefacts illustrated have all been located with a metal detector from sites in the United Kingdom.

A POSITIVE ATTITUDE FROM ARCHAEOLOGY

You may remember that in the Introduction mention was made of the fact that metal detecting, in its earlier years, was not well received by the archaeological community. Largely through ignorance, the hobby was deemed as a threat to our archaeological heritage by way of the destruction of stratified archaeological layers in the soil and the unrecorded removal of artefacts.

The argument that the digging up of finds located by a metal detector causes damage to the stratified layers of soil, which have formed over millennia, has proved to be invalid - owing to the fact that most of Britain's land surface has already been disturbed due to agriculture, deforestation and development. In fact, archaeologists even contradict themselves with this claim by removing the top layers of soil, usually by mechanical means, during the initial stages of any excavation simply because they no longer hold any useful information. Generally, metal detectors only locate objects within these top layers of soil. Rarely do they intrude into the undisturbed archaeological layers below.

However, the argument that many metal detecting finds have gone unrecorded does raise legitimate concerns. Archaeologists now often agree that their initial confrontational attitudes were mostly to blame for the reluctance of many detectorists in coming forward with information concerning their finds.

Fighting a legitimate hobby only proved to be counter-productive. Attitudes simply had to change!

Over recent years, we have seen the hobby become more widely accepted by archaeologists, which has been of enormous benefit to all concerned. Many detectorists now volunteer information by reporting their finds to the relevant authorities. This has helped in building good working relationships between both archaeologists and detectorists. Nowadays, archaeologists quite often solicit the expertise of detectorists during excavations in order to locate and retrieve finds that would otherwise be missed if left to conventional methods of archaeology.

Several years ago, I assisted on the excavation of a Roman site near my own home. I succeeded in recovering many coins, brooches and other artefacts, which would certainly have been missed if it were not for the metal detector. I now gaze upon those artefacts with great pride every time I visit my local museum.

It is an unfortunate, and perhaps sad, irony that the institution that once threatened our hobby has recently come under threat itself. Many archaeologists have suffered from the financial constraints that have been imposed by local authorities. The need to economise has led to job losses and curtailing of resources. Excavations are now only carried out

Fig.9.1. *This fabulous hoard of funerary items was discovered by detectorists in a Hertfordshire field.*

ahead of major construction works - with the developer bearing most of the excavation costs - or when an archaeological discovery is made that needs an urgent rescue dig.

As unfortunate as this situation sounds, it has created a rather unusual paradox. During a recent meeting with my local County Site Recording Officer I was informed that archaeologists are now becoming increasingly dependent on the information provided by metal detectorists in improving their knowledge of the county's archaeological record. A kind of strange symbiosis now seems to be emerging, where both parties work to the mutual advantage of each other.

Many archaeologists now entertain metal detecting clubs across the country by giving talks aimed at improving the detectorist's understanding of their work. The sharing of knowledge and information has now led to a much-improved environment, which has been of enormous benefit to all concerned, as well as enriching our knowledge of the past.

Other changes that have occurred involve the reforming of Treasure Trove, and a voluntary scheme for the reporting of finds. Details of this, and how you go about reporting your finds, are dealt with in a later chapter.

Fig.9.2. *Archaeologists later discovered that a hoard found here was associated with an enormous Roman villa.*

IDENTIFYING, RECORDING, CLEANING, STORING & DISPLAYING FINDS

Identifying Your Finds

The subjects covered in this chapter represent some of the most important aspects of metal detecting. After all, to make such finds in the first place you have put yourself through all the hard work of research, permission seeking, and fieldwalking. There seems little point to any of these activities if you do not identify, clean and conserve what you find.

To tell no one of your finds and leave them unreported, is nearly as bad as leaving your coins and artefacts un-recovered in the ground. A Bronze Age axe, for example, might be 3,500 years old, and in your lifetime you are only going to be its short-term custodian. Having recovered and rescued such a wonderful find, care must be taken to ensure it is not subjected to poor storage, or harsh cleaning.

If you are a newcomer to detecting, and you think you may have made an important find, it is advisable to seek professional advice regarding cleaning and conservation. Some museums will offer such advice, or may even undertake limited amounts of conservation for you. However, if this becomes too regular they may rightfully expect you to donate further finds to their collections.

As a detecting group we are always amazed at the knowledge we have individually accrued over the years. For example, copper alloy - that develops a certain patina - will eventually become familiar in terms of age and colour in your search areas. Some of us have got pretty good at this recognition, and when a fellow Pastfinder shows a coin etc to fellow members, we can state where it was found with reasonable accuracy. When a curious find is made it is now quite unusual that somebody in The Pastfinders does not have a clue as to its identification. How do you get to this stage? Well, thinking back to the past, it just seems to build up over time; however, there are many rich informative sources to consider.

Many museums have identification facilities. Metal detecting colleagues are great pools of information, as undoubtedly they will have read and studied different books to you, and will have conducted research into areas you are not familiar with.

To assist you in the identification of finds we would recommend purchasing Gordon Bailey's excellent "Detector Finds" series (of which five volumes are currently in print). These provide one of the best foundations to classifying finds that one can obtain, and are superb because they cover such a wide range of detector-found items. In addition to this **Treasure Hunting** magazine publishes a monthly "Questions & Answer" section in which readers' finds are identified by experts.

Concerning **Treasure Hunting** magazine, it is important that you keep and safely store all your back copies; in this way you will build up what is probably one of the best ever reference libraries available to detectorists.

Visiting museums and antiquities dealers will also broaden your knowledge; here you will see artefacts and be able to relate to them when you make your own discoveries. Many dealers produce coin lists and mini brochures, which in themselves are well worth keeping.

While many people prefer to detect alone, rather than in groups, with regards to help in identifying finds it is a good idea to join a club. Many clubs hold finds identification evenings, sometimes with a guest specialist. Within a club you will be surrounded by like-minded people, and gain experience from seeing the finds that other members are making.

The coming of the Internet has made the identification of finds a lot easier. There are now hundreds of detecting, coin and antiquity Websites, each showing images of finds and providing helpful hints. Many of the established metal detecting clubs have Websites. Even if you are not a member of the club concerned, most are only too willing to give beginners assistance with finds identification. Such Websites are too numerous to list, but are easily located through search engines and links from one site to another. **Treasure Hunting** magazine also has its own Website: www.treasurehunting.co.uk with a list of UK clubs, detector field tests, and much other useful information.

One recent Website that we tested is worth noting, as it exemplifies how the Internet works. One Sunday afternoon I had the good fortune to find a Celtic silver unit (right next to it was a silver Saxon *sceatta*, but that's another story). I had not a clue as to what the specific identification of this tiny little Celtic coin could be. However, I remembered Chris Rudd's advert in **Treasure Hunting** magazine and within a few minutes sent a scanned image of the coin to him. A few hours later its identification came back. The coin was a North Thames snake and lyre unit, classified as "extremely rare".

Recording Finds

Many of the finds that you make will not be of national importance and will not be worth recording (ie spent shotgun cartridges). However, as you progress in the hobby and start to reap the benefits of your research and fieldwalking, it is likely that you will start to make finds that are of some interest and perhaps value.

Taking Global Positioning readings, and transferring them to a hard copy map or computer database represents one aspect of recording. It could mean taking a find to a museum for recording, and such may indeed be necessary pursuant to The Treasure Act.

Even with items of only average significance you should make some type of record as to where you found them. A growing plot of such finds could lead to the discovery of a habitation site or similar, and this in turn will help identify such areas in future.

Photography can be another very important aspect of recording your finds. This has been made easier with the advent of digital photography and the high definition films now available for single lens reflex (SLR) cameras. Albums with details written on the reverse of each photograph (such as find spot etc) are an invaluable means of recording. This is particularly important should you ever come to part with any finds due to swapping etc, as you will always have a record of them. In the unfortunate event of theft, such an album of photographs could be of great assistance in the possible recovery of the items. This is perhaps a good point to mention the necessity of having adequate insurance when you build up a collection of coins and artefacts. Many dealers will help with valuation, often for no charge, and the value of your collection can be added to your household contents policy.

The keeping of records is not only important now, but will be of assistance to future researchers and historians. In 100 years time your records, however trivial seeming now, might stop a housing estate development smashing through a Saxon settlement or suchlike. Not only that, but finds recording onto maps etc show up areas of intense use and therefore increased loss potential; a year or two on a good site and you will see just where to go to find the most coins etc. The whole idea of local to national reporting is to build up an important record of the history of this island. This is something that as an already responsible detectorist or new starter you should be keen to contribute to.

Cleaning Finds

This is a subject that can, and has, caused much controversy over the years. Fortunately, the attitude is now one of conservation rather than hard cleaning which often removed the patina from artefacts. I once heard of a superb Hadrian *sestertius*, in good condition apart from some slight encrustation, that was later placed into a barrelling machine. Can you even imagine that happening now? Barrelling units grind away surface deposits using pins or bearings and are related to the pebble polishing units. On no account should any ancient or suspected ancient artefact be cleaned with a barrelling unit, as it will strip away the entire patina if any should be present.

Should you wish to read about cleaning and treatments, before putting any into practice, there are several excellent publications available that will serve as good guidelines as to what actions are available for you to take. Many chemicals are now available for the treatment of metals and the various surfaces and encrustations that these can possess.

Alternative methods to chemical cleaning, where appropriate, can range from the previously mentioned barrelling machines and brass wire tipped pencils to the latest ultrasonic cleaning devices. I have also seen some superb cleaning results from using fibreglass tipped pens.

Objects of different metals, if corroded or encrusted, require to be treated in many diverse ways. Items of good quality silver and gold hardly need more than a wash in warm soapy water to remove encrusted soil deposits. However, from ancient times silver has been alloyed with a wide variety of metals including lead, and a low silver content alloy may be referred to as "base silver". Base silver alloys when used for ancient coins etc, become very brittle with age and crystalline in structure. This crystalline characteristic can be very evident on broken or chipped Roman Republican coinage, and to a lesser extent with some of the base silver *denarii* from later reigns (ie Severus Alexander). Some base silver coin cores can be found coated in silver of a higher grade, but this may flake off revealing the duller core. Unfortunately, with old base silver even the best cleaning will produce a coin or artefact that is dull grey in colour. Therefore take great care with cleaning if you suspect your find may be plated. Sometimes you will have a clue as a tiny chip or surface flake may reveal a red copper oxide core.

Several Roman coin types can be found with silver-plating on copper alloy cores; in our experience these have been mainly *denarii* of Vespasian. Whether these are actual forgeries is debatable, as some appear to have been struck from official dies. Some of Henry VIII's coinage was so debased it was produced from copper coin cores and then coated in high-grade silver - hence the name "old red nose" when the coating wore thin. The harsh cleaning of such a coin today would certainly cause "old red nose" to reappear.

Low grade silver can be subjected to various chemical agents to remove the black oxide but this will reduce or totally eliminate any interest such an object might have to a collector. Some silver hammered coins will be found almost dark blue or black with oxide. Where the coins are not valuable but still worth enhancing (carefully check first!), we have found that hot water and toothpaste will remove the oxide. By gently rubbing the toothpaste over the surfaces, much of the oxide will be reduced. But remember toothpaste is a mild abrasive so, as stated before, do not use on rare or valuable silver coins! Some hammered silver coins respond excellently to being wrapped in tin foil and rubbed, a reaction will occur resulting in mild heat generation and a cleaner coin.

Copper alloy objects require more careful consideration, for which a variety of waxes, lacquers and bronze disease treatments are commercially available. Some ancient copper alloy coins such as Roman *sestertii* and *dupondii* can be found with a very thin slate-like smooth encrustation on their surfaces. We have found that gentle tapping with a wooden spoon or similar can dislodged this, often revealing a beautiful patina beneath.

Where this type of encrustation has crept into lettering and hair detail on a bust, you can remove this with the extremely delicate use of a scalpel blade or a needle viewing the object through a microscope or similar. However, we do stress that this should be only attempted on objects of no historical or financial value. You need an extremely steady hand for this, not only to safeguard the coin, but also your fingers!

Some copper alloy artefacts can be coated in a thin layer of "silvering" for decorative purposes or to give the illusion of being silver. In most ancient examples the metal used is tin, and some Roman brooches you see or find will be referred to as being "tinned". If you suspect this to be the case with something you have unearthed, seek expert advice before any chemical cleaning. Some cleaners and chemicals will strip this tinning off within seconds, therefore ruining the item.

Sometimes you may discover iron objects that you wish to preserve and keep. Unfortunately, these can be among the most difficult items to conserve. This is because iron reaches a certain stability within the soil; as soon as you remove it, it becomes

Fig.10.1. Silver shilling of George III as found.

Fig.10.2. The same coin after cleaning with citric acid. (Never use abrasive materials).

unstable. This can result in severe cracking and flaking and sometimes the exuding of an orange viscous fluid. There are some chemical treatments available, but with seriously degrading artefacts there is little you can do. We have found in these cases that soaking in distilled water, drying in an oven, cooling, and then liberally dosing with one of the powerful contact adhesives is about the only option. At least with this method, the decay of a seriously unstable iron object is halted, and its shape retained.

With all such objects, if you have the slightest doubts then refer and leave any cleaning to professionally qualified people. You should definitely not attempt to clean any hoards of ancient coins or artefacts as ideally they should be cleaned after discovery by professional bodies. In fact, by actually cleaning certain categories of items you may well end up in breach of the Treasure Act.

There are vast arrays of cleaning agents available, although some are very specialist. However, some can be dangerous - both to user and finds - and the instructions supplied should be strictly adhered to.

All of these chemicals are widely available from reputable detector accessory suppliers. Please note, however, that in all cases pre-trial testing should be carried out on an unwanted "tester" artefact of a similar metal. As in all cases, results can never be guaranteed. Should you unusually witness a reaction that is obviously destructive, carefully remove the artefact as soon as possible without risk to yourself. Further to this, read all instructions fully and never mix chemicals unless directed. The importance of wearing rubber gloves, eye protection, and chemical applications being used in fully ventilated areas cannot be overstressed.

The following is a basic list of some of the cleaning agents available, their actions and the metallic surfaces they should be applied to.

Sodium Hexametaphosphate

Dependent on the level of encrustation a five minute immersion in this chemical will remove most minor encrustations on copper alloy objects.

Sodium Sesquicarbonate

This chemical is very useful for removing chloride deposits from iron, as well as copper and its associated alloys.

Bensotriazole

This is one of the most popular treatment applications for bronze disease.

Incralac

This acts as a protective treatment, mostly used on copper alloys.

Microcrystalline Wax

This treatment can be very effective on the surface of ancient dry chipped coins. The wax requires heating before application to the item. It can cause a mild colour change to some patinas and dries with a matt finish. Large objects may require several applications, and some coins or artefacts may continually require further applications to keep up their appearance. Some detectorists coat coins with patina in black boot polish. This does enhance and define features to a higher degree, but again this effect fades with age. In our experience microcrystalline wax is vastly superior.

EDTA

An extremely useful application for the treatment of iron corrosion.

Storage & Display Of Finds

Having taken the time and trouble to locate your finds, you should make adequate arrangements for their safe storage. I have known of some good finds being stored in plastic bags for years along with general rubbish items, and items being stored all over the house in small piles. This is certainly not only guaranteed to irritate other family members, but also to end up with important finds being lost.

No person has a right to extract the history of this land, and then subject it to more damage than it has potentially succumbed to in the last hundred or thousands of years. As a group, we have to admit that nothing annoys us more than when we hear of such finds mismanagement. Seeing once superb, but battered Roman coins extracted from a bag of other unwanted finds is irresponsible and unacceptable.

Some excellent coin and artefact trays, made of high quality plastics, are commercially available. These have the added advantage of giving a very organised and presentable appearance to your collection. After all, you should be very proud of your achievements.

Making your own finds display cabinets is another option. However, if you are using wood in their construction please seek expert advice regarding your intentions. Woods such as cedar can leak harmful resins and other side products such as vapours. Some woods can cause serious corrosion and discoloration to metallic items. Another wood to avoid is oak. Among the suitable woods are: mahogany, rosewood, and walnut. Always store your finds in dry conditions and away from direct sunlight.

Dependent on the moisture level of your storage area, it is possible that after cleaning you may find that some of your Roman silver or medieval hammered coins start to go darker again as they re-oxidise. Excessive handling, due to acids and moisture from our fingertips, can also increase this. Some people might disagree with this preservation method, but we apply a thin coating of plastic lacquer to all our silver coins. You can then handle them as much as you like, and they stay wonderfully clean in appearance. Should any perfectionist wish to remove the lacquer in the future, then warm soapy

water and an old soft toothbrush will achieve this admirably. Lacquer does not chemically harm silver coins and serves as a protective barrier.

One further option for coins is to store them in an album; however, there are a number of issues to consider if doing so. If storing in simple album format take care that heavy coins do not damage coins stored on adjacent pages, when flipping through the album. On large coins with a good patina this could cause chipping to occur, creating small deposits of greenish powder at the base of the coin pocket.

Always ensure that the album and the sleeves you obtain are from a reputable manufacturer, as several of the plastics used in the production of some albums can release chemicals and vapours that could be harmful to coins over a period of time. Storage in a cool low moisture area is suggested to prevent the plastic sleeves and pockets becoming subject to heavy condensation.

One very useful storage and display idea that we have seen is to place a small plastic laminated piece of paper with each find. On each of these is written what the item is, the date of discovery, and the map reference of the find. For coins this identification ticket is placed underneath, for larger finds it can be tagged to them with cotton or suchlike, creating a very informative and visually effective display.

Disposal Of Surplus Finds

As your detecting career progresses, it is probable that you will accumulate several coins or artefacts of exactly the same type. The question is then, what should you do with them? You could of course keep them, in the hope that you might be able at some point in the future to swap them with fellow detectorists for items that you do not already have in your collection. Alternatively, you could approach local schools or museums to see if their history departments or displays would like such finds.

Another course of action is to approach an antiquities dealer with your finds, with a view to swapping them to fill gaps in your collection or for financial return.

Some dealers will also purchase bags of broken finds, which is also very good for the hobby. The result of this is that there are lots of "partefacts" in circulation that are ideal for junior collectors as they are affordable due to their condition.

Be prudent whenever classifying any find as rubbish, because you can discover later that there is someone out there who collects what is your "rubbish". Old rifle shell cases and Martini-Henri type bullets, dismissed by many, can be very collectable to others. What may be junk to you can be truly treasure to another.

Don't forget to regularly check your scrap bags, because as your knowledge increases so does your ability to recognise artefacts. What was considered unidentifiable junk last year may actually be a fragment of a rare Saxon brooch.

Inevitably you will find many unwanted metal items during your searches. All such junk should be removed from the land (why not show the landowner just how much toxic metal you are removing from his land, as yet another advantage of having you detect there?). This scrap removal will also cleanse the site, so that at least you will not re-find such rubbish next year. Unwanted scraps of metal should be taken home, wrapped up safely and disposed of through your local waste authority services. Metals such as lead, copper, bronze and aluminium may, if saved in quantity, have a small commercial value from scrap metal merchants. In carrying this out you are engaging in the important process of recycling, as well as making some battery replacement money.

MAKING A CONNECTION WITH YOUR FINDS & CREATING "TIME LINES"

One of the most interesting aspects of this hobby is connecting with the finds you make. With very ancient artefacts this can truly stretch your imagination. You might find yourself asking questions such as: "What was the name of the person who last held this little Saxon *sceatta*?" or "What did he or she look like?" Perhaps you even create your own answers and write them in a log of your finds.

Although marvellous and mind stretching, historically such mental wanderings can never be factually based. What a shame that you will never meet the actual owner of that superb Roman enamelled brooch you have just uncovered. An even greater shame perhaps is that you are unable to meet that medieval moneyer who kept miss-striking those short cross pennies you find. Despite this, by simply finding the item you create a bridge across time connecting you to the person who lost the item. This is what we term a "Time Line".

Technology - it is argued - may never be able to produce a functional time machine, but until it does the metal detector admirably fills the gap. However, there are finds of more recent times where you not only create a time line to a past loss but, if you are very lucky, you may meet the very person who experienced the loss, or direct members of his/her family. Examples of this are where wedding rings and bracelets etc are returned to their owners. I read recently of a named charm bracelet returned to its owner who lost it over 50 years before. You only have to read **Treasure Hunting** to see how many rings and other treasured items of jewellery are found and returned to their very relieved owners.

Other examples of fairly modern losses send you on research projects that even a week before you may have had no particular interest in (ie the frequently found military buttons and cap badges). Metal detecting has created some phenomenal experiences for us, that we feel privileged to have been involved in. For example, some years ago I was involved in organising the reunion of a Luftwaffe Flight Mechanic and the very RAF pilot who had shot him down in 1941. He had been part of the crew of a Junkers 88 on a raid to Birmingham in 1941 when he was intercepted. We learned later that, ironically, the RAF pilot had by shooting down the German bomber actually saved the life of its crew. This was due to the Flight Mechanic's Staffel later being posted to the Eastern Front. No one came back alive from that venture.

You can imagine how emotional this meeting was, as they proclaimed each other "brothers of the air". It was made all the more remarkable for me as, due to an earlier metal detecting search of the spot where the Junkers crashed, I was able to hand back to the flight mechanic one of the spanners from his on board tool kit.

Another incident springs to mind when I was metal detecting the crash of a Heinkel He 111 bomber shot down in 1940. Finding lots of interesting pieces, I later sent some to the original pilot now living in Canada. This pilot sends me a Christmas card each year. Every time I read a book on the Battle of Britain and see him mentioned, it means more than just a name in the text.

It gives you such a sense of achievement to be involved at this level, and to be able to do such things is a true honour. I will never forget the time when a Mosquito pilot was telling me about the time that the whole starboard wing of his aircraft suddenly broke away, and he was literally sucked out of the damaged cockpit. "The last thing I saw, boy, was the altimeter needle going haywire". Later we organised a small excavation of the site where his aircraft had crashed. Fortunately, the pilot was able to attend, and was fascinated with the day's events. A lot of luck and the use of a metal detector allowed me to present him later that day with the very same - although somewhat battered - altimeter. You simply should have seen the look on his face. Sadly, the young navigator was killed in this crash, and as a mark of respect we asked the local vicar to hold a service at the site of the excavation. This was duly done, and at the end of the day a sapling oak tree was planted exactly on the impact spot.

Once I found a tiny piece of brass inscribed with the name "Lady Bearstead", and an associated address. Photographing the artefact, and later including in it in a **Treasure Hunting** article led to additional research and the later receipt of an interesting letter. This letter was from an elderly gentleman who actually remembered Lady Bearstead in the 1920s, and went on to give me a full Bearstead family history. It would seem Lady Bearstead's husband was a wise investor. In the 1890s he bought virtually all the shares in a small petroleum company and reinvested further; the company did rather well, and today is known as Shell.

We have chosen two metal detector finds of reasonably modern date, which we feel illustrate time lines and often fascinating research that is associated with them. This is, of course, combined with the exciting possibility of meeting or corresponding with the owner or relatives.

Find 1

Fig.11.1. The plaque was folded in half when first found.

While searching in an isolated field in the heart of Bedfordshire, I unearthed a very large bronze disc. The disc was heavily corroded and had been bent in half, presumably from being struck by a ploughshare. Although seemingly uninteresting, I kept the disc and took it home with me, where it was consigned to a box containing many other items of metal "junk" that I had accumulated over recent years.

A year or so later, while sorting out this "junk", I picked out the disc and almost threw it away. Fortunately, I had the sense to examine the disc more closely before doing so. Holding it up to the light, I began to make out details of a figure - most of which was obscured as a result of the disc being bent double. I decided to make an attempt at straightening out the disc in order to try and see what the figure

Fig.11.2. A similar plaque in the Luton Museum.

actually was. After doing so, I soon found that it was, in fact, a large plaque depicting a classical figure standing beside a lion. I had no idea what the plaque or medallion represented but I was soon enlightened after a visit to a local museum. In a glass case, along with many other medals and plaques, was one identical to that which I had found. I noticed, also, that the plaque bore the name of a soldier who had been killed in the First World War. When I later returned home I examined my own plaque to see if it, too, bore any inscription. After some careful cleaning I could make out a man's name. The disc now took on a whole new meaning. No longer was it a piece of corroded and twisted "junk"- it now represented a man's life.

Searching through a list of names of soldiers who had died in the Great War, compiled onto a CD Rom, as well as visiting two Internet Websites, revealed an astonishing story....with a puzzling twist in its tail. The soldier's name did appear on the compiled list, along with his Regiment, serial number and the day he was killed. However, when cross-matched with names listed with the War Graves Commission, I soon discovered that the name on the plaque wasn't the man's real name. Now armed with the soldier's real identity, I was able to learn that he had been a greengrocer who lived at Carshalton, in Surrey, before enlisting in the army. His surviving relatives continued living at his home address until the 1960s, but no further trace could be found. I also learned, from reading the Regimental Diaries, that he was one of three men who had been killed while searching a building in Arras, in 1917. The building was hit by artillery fire. For many weeks, this man's story had become a part of my life. I had managed to find out who he was, where he lived, and how he died. The two puzzles I have yet to solve. Why did he change his name when he joined the army? And perhaps even more intriguing, how did his memorial plaque end up being buried in an isolated field in Bedfordshire 40 or 50 miles from where he and his family had lived?

Find 2

Fig.11.3. The identity tag of Lester B. Culp, who died on the "Peacemaker" B17 bomber.

For many years I have been fascinated with the crash of several B17 bombers around the village of Weston in Hertfordshire. One of these which was named "The Peacemaker", crashed on a training flight in 1945. Having obtained the crew names, somebody suggested I use one of the name-tracing Websites, to see if any relatives still lived in the USA.

Basically, all I had was a surname and place of birth. Would anyone still be there after 60 years? Perhaps the family had died out or moved away. The Website revealed 15 people with the same surname, so I sent a mail shot to them all. Nothing happened for a few weeks and then a solitary letter arrived. Amazingly it was from the younger brother of one of The Peacemaker's crew, all of whom had been tragically killed. He later went on to send me press cuttings, family photographs and a whole history of family events.

For many years I have metal detected where The Peacemaker crashed, collecting some small twisted fragments and several bullet cases. Then one day another detectorist appeared on the site, also passionately interested in such wartime happenings. We had a good old chat and then carried on our separate searches

A few weeks later I received a telephone call from him stating he had found something very unusual. Later we met up and unusual his find certainly was: he had found the dog tag of the crewmember whose family I had traced in the States. I managed to persuade the gentleman that, in the light of the fact that I was already in contact with relatives, we should arrange to return the dog tag. This was duly done, and resulted in one of the most passionate thank you letters I will ever receive.

One off shoot of this find, was that through correspondence with the brother I discovered that the hometown of his family in the States is famous for producing pretzels. It was my turn to be surprised a month or so after when a huge 10k presentation of the aforementioned snacks was delivered by courier to my doorstep. As a young boy of approximately 10 when I first heard of this B17 crash, I could never have believed that in three decades time I would be in contact with the family of one of the crew.

FURTHER READING & ASSISTANCE

There is a vast amount of published work available on metal detecting, and all its associated interests and specialised subjects. Some of these can be very reasonably priced, while more specialised work can often be quite expensive. To pay over £100 is not unusual for some reference works on Roman coins, for example.

The development of your attitude towards metal detecting, and the areas of interest you adopt, will largely influence what reading sources you choose. However, finding an unusual button or 17th century trade token, may see you searching out and buying works on subjects that a few days ago you would never have considered.

That is one of the great things about metal detecting....you don't have a clue what your next find will be or what information you will next be researching as a result of that find. Because of this natural and individual variation, it is impossible to cover all helpful publications in the list below.

Recently, another piece of assistance to the beginner detectorist has arrived and this is the video. Several are available; however, the one that springs to mind is Mike Pegg's production called "Metal Detecting Down to Earth". Mike's enthusiasm is difficult to beat and emanates strongly from this video.

With regards to television there are two programs that are extremely interesting, as they show the superb results of responsible detecting coupled with research. These are "Hidden Treasures" and "Two Men in a Trench".

The following, however, are the literary publications The Pastfinders consider to be among the core requirements for a general overview of finds made in the UK and essential to the detectorist's library. In addition to our recommendations, though, please remember there are numerous other published works that are also excellent.

Benets, Artefacts Of England & The United Kingdom

Compiled by Paul Murawski and now in its second edition, this is one of the most luxurious of metal detecting orientated publications. It covers finds from the Stone Age to the medieval period. Issued in hardback, the quality of the illustrations is superb. This publication will never be found far away from any serious detectorist. One or two of the illustrations are actual Pastfinder discoveries, and we are delighted that they have been included.

Nigel Mills

This author and antiquity expert has produced three very valuable aids to artefact/antiquity identification, resulting from his unparalleled experience of detector finds. The three volumes currently in print are: **Celtic & Roman Artefacts**, **Medieval Artefacts**, and **Saxon & Viking Artefacts**. As in so many hobby-associated publications, it is not just detectorists who are using these books as identification aids. (Greenlight Publications).

Gordon Bailey's Detector Finds Series

At present there are five individual volumes in print, with hopefully more to follow. This series is really the backbone as regards identifying some of the more common and rarer finds that you may encounter. Have you found a strange fragment of something, or an unusual small artefact? The chances are that you will find it illustrated in these volumes. As is often the case in hobby literature, these books have been written by a detectorist for detectorists. (Greenlight Publications).

Brian Read's Metal Artefacts Of Antiquity

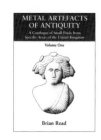

An excellent publication dealing with small metallic finds from specific areas of the country but relevant to the whole of the United Kingdom. Hundreds of archaeological-standard line illustrations by Patrick Read and Nick Griffiths.

Roman Coins Found In Britain

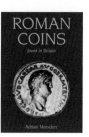

Written by Adrian Marsden, this has to be one of the most detailed and thorough works on the vast subject yet published. It includes great detail on the coins and many photographs associated with mint-marks. How many of us are good at interpreting these on Roman coins? The book also comes with a separate up-to-date price guide. (Greenlight Publishing).

Shire Publications

There is an excellent section of archaeologically-based subjects covered by this series of pocket-sized books. These topics range from later prehistoric pottery to medieval town plans, and are dealt with in superb detail. Museums and most good book shops normally stock a large selection, and a free list of all titles can be obtained from the publishers.

Spinks Coins Of England & The United Kingdom

This is an unbeatable book for all those who admire, find and collect coins. Beginning with a detailed section on Celtic coins, it ranges right up until the latest issues of Queen Elizabeth II. It is the most comprehensive overall single volume available on UK coinage, and is considered to be a "must" by most detectorists.

Hidden Treasure

This is the recent publication brought out to accompany the above-mentioned BBC TV series. It is a well-illustrated work, showing the superb results of responsible detecting in the UK. The series and book largely concern themselves with magnificent metal detecting finds such as the Ashwell and Wheathampstead hoards, and the magnificent Ringlemere gold cup. It contains little coverage of the day-to-day basic finds we all make, but then again we all live in hope don't we?

Dealers Lists/Pamphlets

Many dealers produce lists, catalogues etc either in hard copy or available on mailing lists via the Internet. Collecting and indexing these will undoubtedly assist in coin and artefact identification.

Auction House Catalogues

Some of these are lavish publications in their own right, packed with colour photographs on glossy paper. They can be rather expensive to subscribe to, but are worth acquiring. They provide yet another very useful reference source.

FURTHER INFORMATION

The information given in this chapter could, in some cases, be regarded purely as common sense; and some facts might be a repeat of what we have already mentioned in previous chapters. However, we felt it would be beneficial to condense this information, besides providing some more details.

Some of the following subjects are particularly relevant in relation to conduct, and may not be so "obvious" to a beginner.

In the UK there are many people, such as members of FID and the NCMD, who have worked hard to obtain and help maintain the respectability that metal detecting has today. Therefore, with respect to such person's efforts it is essential that no one detracts from this, as setbacks could take years to rectify.

Metal detecting is not the simple pastime it was 20 years ago. However, this loss of simplicity has brought great benefits for metal detectorists and archaeologists alike. There are now important laws and schemes that are designed to increase our knowledge of the past while also protecting our heritage. The primary examples are The Treasure Act and The Portable Antiquities Scheme.

The Treasure Act 1996

This important Act came into force on 24 September 1997, and relates to finds made in England, Wales and Northern Ireland. The Department for Culture, Media and Sport (Tel. 0171 211620) have issued a leaflet called "The Treasure Act, Information for Finders of Treasure (England and Wales)"; a separate version is also available for Northern Ireland.

This extremely informative leaflet covers situations such as: What is treasure? And what do you do if you suspect you may have found some. Further to this, and also from the same department, can be obtained the more detailed "Code of Practice on the Treasure Act". Again, a separate Code of Practice is obtainable for Northern Ireland.

Both documents are invaluable to today's detectorist. Being prepared for that find of a lifetime is a good idea, and we recommend obtaining both leaflets.

Basically, the Act is concerned with artefacts and coins of gold and silver that are at least 300 years old when discovered, but greater details will be found in the Code of Practice. Since 1 January 2003 the Act also encompasses Prehistoric metal assemblages. The primary source in the UK for this new category would be Bronze Age hoards.

It has to be said there have been criticisms in relation to this Act, primarily the length of time to process and value potential Treasure finds. However, most new systems need to be refined in some way or another, and it is hoped that with time these faults will be highlighted and rectified.

The Pastfinders have had some experience of the Treasure Act, in relation to the declaration of a tiny 17th century gold mourning ring. Our experience was one of efficiency and good responses from the museum curator, the local media, and the court hearing with the Coroner.

The Portable Antiquities Scheme

This Scheme came into being in 1997, and is really a complimentary measure introduced in support of the Treasure Act. It concerns the voluntary recording of finds of an archaeological nature located by members of the general public. Such finds could encompass anything from a Palaeolithic flint hand axe to a single gold coin from the reign of Elizabeth I.

To monitor this scheme the position of area Finds Liaison Officers was created, and is now very successfully established. Members of the public can now either contact their local museum, or the FLO directly to report their find. This is a good idea, as each year hundreds of thousands of items are found that do not qualify as treasure and previously went unrecorded. The scheme is designed to increase our knowledge and, combined with the Treasure Act, has certainly proved successful.

In most years about 98% of all recovered metallic artefacts are located with a metal detector. Both the Treasure Act and Portable Antiquities Scheme show that at last metal detecting has been seen to be the most valuable source of metallic artefact location possible. Having finally won the recognition that this hobby and all those involved so richly deserve, the response from metal detectorists has accordingly been phenomenal. Today you will find that most metal detecting clubs, groups and individuals are now fully involved with the PAS, and the story of its success in information recording becomes richer in detail and importance as a result.

More details of the Portable Antiquities Scheme can be found on the associated Website www.finds.org.uk

Treasure Trove Laws In Scotland

The present legislation applying to the reporting of finds in Scotland is substantially different from that applying to the rest of the UK. In Scotland many finds of historical or archaeological interest, whether of precious metal or not, need to be reported and may be classed as Treasure Trove. If you live in Scotland, and would like further information and literature contact:-

National Museums Of Scotland
Chambers Street
Edinburgh EH1 1JF
Scotland
Tel: 0131 47 4422
Fax: 0131 220 4819
Email: info@nms.ac.uk
Website: www.nms.ac.uk

USEFUL CONTACTS

National Council For Metal Detecting
Trevor Austin
NCMD General Secretary
51 Hilltop Gardens
Denaby
Doncaster
DN12 4SA
Email: Trevor.Austin@ncmd.co.uk
Website: www.ncmd.co.uk
(Membership includes Third Party liability insurance. Copies of Landowner/Searcher agreement forms available to members).

Federation Of Independent Detectorists
Colin Hanson
"Detector Lodge"
44 Heol Dulais
Birchgrove
Swansea
West Glamorgan
5A7 9LT
Website: www.detectorists.net
(Membership includes Third Party liability insurance. Copies of Landowner/Searcher agreement forms available to members).

British Museum
The British Museum
Great Russell Street
London
WC1B 3DG
Tel: 020 7323 8611/8618
Fax: 020 7323 8985
http://www.britishmuseum.co.uk

National Museum Of Wales
Cathays Park
Cardiff
CF10 3NP
Tel: 029 2039 7951
Fax: 029 2057 3321
Website: www.nmgw.ac.uk

National Museums Of Scotland
Chambers Street
Edinburgh EH1 1JF
Scotland
Tel: (+44) 131 47 4422
Fax: (+44) 131 220 4819
Email: info@nms.ac.uk
Website: www.nms.ac.uk

Finds Liaison Officers
For details of your local Finds Liaison Officer (or for more details about the Portable Antiquities Scheme) write to:
Portable Antiquities Scheme
C/o Department of Coins and Medals
British Museum
London
WC1 3DG
Tel: 020 7323 8611/8618
Email: info@finds.org.uk
Website: www.finds.org.uk

Celtic Coin Index
Dr. Philip de Jersey
Institute of Archaeology
36 Beaumont Street
Oxford
OX1 2PG
Tel: 01865 278240
Fax: 01865 278254
Website: http://units.ox.ac.uk/departments/archaeology/ccindex/ccindex.htm

Port Of London Authority
Bakers Hall
7 Harp Lane, London EC3R 6LB
Tel: +44 (0)20 7743 7900
Fax: +44 (0)20 7743 7999
Website: www.portoflondon.co.uk
(Contact for search permits to search the Thames foreshore).

Local Clubs
A full list of metal detecting clubs in the UK is available from the National Council for Metal Detecting and the Federation of Independent Detectorists.

Useful Tips

● If you are searching tidal areas always obtain a Tide Chart to check on Low or High tide times. Some charts also carry information as to any areas of "quicksand". As in all risky locations it is best to search with a colleague, or at least ensure that you have a fully charged mobile phone in case you get into trouble.

● Always obey any legal signs related to the ownership of search areas, and do not interfere or make yourself a nuisance in connection with other people's pastimes on the land.

● Consider taking out an insurance policy that covers you against any possible claims, such as accidental property damage and any ensuing legal costs. Such happenings with metal detecting are extremely rare, but you never know. Membership of organisations such as FID (Federation of Independent Detectorists) or the NCMD (National Council for Metal Detecting) comes with an organised insurance for you upon joining. It is also worth including any valuable coins and artefacts you might have found on your household policy.

● If you take along a packed lunch when out searching, recover all associated litter and dispose of thoughtfully. Should you encounter any litter that somebody not quite as thoughtful as you has dumped, then also please dispose of this as well. Plastic bags, yoghurt style pots, bottles etc are harmful to livestock as well as wildlife.

● Always keep a simple medical kit in the car; you never know when you might encounter an awkward blackthorn or sharp flint. This kit should also contain insect repellent and bite/sting treatment.

● Check all fields before entering for any livestock that may be boisterous or dangerous, particularly if you have the additional responsibility of young children with you. Bulls still occasionally kill people in this country. Young heifers and bullocks can be very curious and extremely playful, and although meaning no harm are nevertheless very heavy. Curious horses have also trodden on many a search head, so beware.
In relation to livestock always close any gates that you open. Should that prize pedigree Hereford bull get into that field of pedigree Frisian cows, then later on you could have some genetic based questions to answer that may additionally involve your own parentage also being questioned!

● In your enthusiasm to get out into the fields don't park in gateways or other field access points, particularly at harvest time.

● If you recover any recently detached or broken parts of agricultural equipment when out detecting, always return them to the landowner or farmer. It's a good idea to have a prearranged placing area, so you can just put any items there to await collection.

● Soil contains many harmful varieties of bacteria and some of these are potentially lethal. Never allow soil covered fingers to come in contact with your mouth (eg when eating sandwiches, smoking a cigarette etc). A bottle of tap water and some medicated soap is easy enough to take out with you. Try to wear gloves at all times; the thin plastic surgical variety are ideal. Fields that used to contain horses represent a risk from anthrax infection, so ensure all cuts are covered with a plaster before and during searching.

● Never put yourself or the search head of your detector near to an electric fence. If you make contact with it you will receive quite a "packet", but will survive. Your metal detector, however, will probably blow all its circuitry and will not. Such fences normally consist of single or multiple wires (sometimes yellow in colour), attached to supports with conspicuous rubber cotton reel like fittings. When it is misty or has recently rained these fences often click or pulse quite loudly. Another good indication is that your metal detector may receive interference interspersed with a series of regular pulses. If in doubt ask the landowner.

● If you are checking marshy areas always tread cautiously, and preferably go with a colleague. If in any doubt call off your search - the risks are never worth taking. Other areas to be very wary of are farmyard or local effluent pools. These can dry with a crust on them, appearing safe to cross over, but below the crust can be several feet or more of liquid matter. Also in relation to marshy wet areas, particularly in wintertime, keep a spare pair of socks in the car. You never know when in lazy mode, or over-eager anticipation, you misjudge the width of a water filled ditch.

● Eroded riverbanks can be very hazardous. Always try to gain a preview of the bank condition from the other side of the river.

● Adders seem to be on the increase. In summer use caution when searching sand dunes, heath-land and any area with dense dry grass. If bitten, medical attention should be sought as soon as possible. To a healthy adult the bite is usually very uncomfortable but not fatal, but young children and the elderly may have a severe reaction. In known high-risk areas there will often be adder warning notices.

● Make every attempt not to disturb local wildlife. Wild birds venture far closer than normal when you are too near their eggs or newly fledged young. It is wise to stay away from any woodland where the gamekeeper has stock breeding pens, and where wild pheasants will be nesting.

● Should you encounter quantities of live ammunition or a bomb, flare, large shell or rocket, then clearly mark the area, and withdraw immediately. Contact the landowner as soon as possible, and the police, who in turn will contact Bomb Disposal. Never attempt to recover the item, move it, or test it with your spade, as many old explosives are highly unstable and are particularly susceptible to vibration and or temperature changes. On high explosive ordnance of some age, sometimes the inner filling will degrade, particularly when the casing is damaged, and may appear on the surfaces as a highly dangerous white crystalline deposit. If you are detecting in a high-risk area, it is a very good idea to try and familiarise yourself with some of the ordnance you might encounter.

● Always take a set of spare batteries or an additional re-chargeable pack out with you. Nothing is more frustrating than having forgot to check your power supply status, and later your batteries suddenly expire. While on the subject of batteries, it is advisable not to mix manufacturer types or part discharged with fresh, as this can reduce your detector's performance. If you are not going to search frequently then remove the batteries after use to avoid damage from leakage. Never insert batteries into your detector that show signs of leakage such as white or bluish crystalline deposits, as these can damage the circuitry surrounding the battery compartment. Always check battery condition.

● Binoculars are very handy for checking site conditions, particularly where a long walk may otherwise be involved.

● If you are able to, always try to take a camera out and about with you. Apart from some of the incredible wildlife opportunities you may encounter, imagine not having a camera when all those gold nobles finally spill out from the side of the hole you have just dug.

● If possible take a spare pair of headphones with you. Should the ones you normally use become faulty, this will avoid you creating excess noise by searching without headphones. In summer time it is a good idea to take a spare pair of those small portable CD player type headphones out with you, as the normal large headphone ear cups can make you sweat uncomfortably.

● If you are detecting or fieldwalking in hot weather, particularly on fresh cut stubble where temperatures can become very hot, ensure that you have adequate supplies of fluid to drink, as well as appropriate sun protection.

● A roll of high impact adhesive tape can be very handy for conducting simple temporary repairs to your detector while out in the field.

● Wherever you are, and whatever the surface you are detecting upon, always fill in your holes. This may be an "old chestnut" but so many sites are now not available due to holes being poorly filled in or not at all. One bit of carelessness can cost you a hard worked for site, and perhaps will deny access to future detectorists for many years to come.

● Finally, you never know when you will make that extra special find. We have found it useful to carry separate seal top bags, or small foam-filled plastic containers to isolate and protect such finds.

We hope that you will find some of these facts interesting and helpful, and that they may assist you in starting or continuing to enjoy metal detecting.

GLOSSARY OF TERMS

Alkaline - Non-rechargeable type of battery, recommended as being the best to use by most detector manufacturers.

All-Metal - A setting available on some detectors allowing them to register any metal type, whether iron or non-ferrous. On motion detectors this is often a non-motion mode used for pinpointing targets or for manual ground balancing.

In Air Test - A test performed by passing various sample targets over the detector's search coil with the detector resting on a table or bench. Although useful for checking discrimination settings etc, depths achieved during such tests do not accurately represent how the detector will perform when locating objects in the ground.

Artefacts - Man-made objects usually of some age. In relation to the hobby of metal detecting the term usually refers to buckles, buttons, musket balls etc as opposed to coins.

Audio I.D. - The facility available on some detectors to identify targets by audio tone. Usually the tone will drop for iron or small junk items (silver paper etc) and rise when wanted targets are encountered.

Bench Test - (see In Air Test).

Black Sand - Negative ground minerals consisting of magnetite or magnetic iron oxide. Present on both inland sites and beaches, this form of mineralisation can cause great loss of depth from a metal detector or in extreme cases make a site unworkable.

Body Mount - A detector with a control box that can be detached from the stem (or is made to only work in such a configuration) and mounted on the waist belt or a chest harness to save weight and reduce arm fatigue.

Chatter - Broken signals or noise caused by ground minerals or the presence of small pieces of iron. Can usually be overcome by reducing sensitivity and/or slightly increasing discrimination.

Coil - The part of a detector that puts out an electromagnetic field and senses the presence of metal. Also known as the search head or loop.

Coin Depth - The facility of some detectors to give an indication of the depth of a target by means of a meter read-out.

Concentric - Search head with circular coils of a different diameter tuned to each other. Such coils produce a cone-shaped search matrix.

Control Box - The part of the detector housing the electronic circuitry, controls, and sometimes battery compartment and meter.

Controls - The switches or touch pads that allow a detector to be switched on and adjusted by the user to the required setting.

Cut-Back - Loss of depth caused by mineralised soil or the salt-wet sand of tidal beaches.

Depth - The distance, measured from the surface, at which a detector has registered a find.

Detectorist - Colloquial term used to describe a person who uses a metal detector.

Digger - Trowel or "cut-down" spade used to recover finds. Most normal garden types of trowel are not usually strong enough for metal detecting. Digging tools have been modified particularly for hobby uses and are readily available, the best being those made from stainless steel.

Discrimination - The ability of a metal detector to ignore unwanted items such as iron nails, silver paper, or ring pulls. In most circumstances experienced detectorists keep this set to no higher than the "Iron" level as there is a chance that wanted items may be rejected along with the junk.

Double-D - Type of search head containing two balanced D-shaped coils. Also known as "widescan".

Drift - Where a detector moves from its preset threshold point as a result of changes in ground mineralisation, external temperature changes or instability within its electronic circuitry. Most modern detectors either do not drift or have in-built auto-tuning that automatically adjusts for any drift.

Eyes-Only - A form of searching normally carried out without the use of a metal detector. It is usually carried out on freshly ploughed/harrowed fields or after rain when objects are easier to spot on the surface.

False Signals - Signals that appear to come from a wanted target but when re-checked have "disappeared". They usually result from junk, a loose coil lead, or the search head hitting stiff stubble or a clump of hard soil.

F.L.O. - Finds Liaison Officer. Appointed official for a county or area, normally a qualified archaeologist, in charge of recording and identifying finds made by the public under the Portable Antiquities Scheme.

Ferrous - Objects made from iron.

Fixed Discrimination - Normally applying to the older type of detectors, it is where the reject level has been factory pre-set usually to discriminate against iron and small pieces of silver paper.

Foot-Assisted - Long handled trowel with foot bar to allow easier ground penetration.

Frequency - Hertz measurement of alternating current cycles per second produced by a metal detector's transmit oscillator. Most modern detectors work on low or multi-frequency.

Ground Balance - A point at which a detector will neutralise ground minerals and have a better ability to register wanted targets.

Ground Effect - Loss of depth caused by positive or negative mineralisation.

Ground Reject - The ability of a detector to ignore or partially overcome ground effect.

"Grot" - Worn or badly corroded bronze or copper coins, normally Roman or Georgian.

Hammered - A coin that had been struck by hand between dies from a metal coin blank. Normally silver or gold, some low-value hammered coins were struck from copper or copper-alloy blanks.

Hoard - Normally a collection of coins or precious metal items buried for safekeeping, but for some reason never retrieved. The term can also apply to ancient bronze axe heads or broken fragments of same (ie a founder's hoard).

Hot Rock - A rock that contains a higher concentration of detector-reactive minerals than its surrounding area. Will cause most detectors to signal in discrimination mode or "null" (go silent) in all metal mode.

IB/TR - Induction Balance/Transmit Receive. Most modern detectors work on this principle.

"Iffy" Signal - A broken and sometimes "one-way" signal usually caused by a small wanted target or one at maximum depth range. Although sometimes resulting from a piece of iron or similar, all such signals should be dug up.

Junk - Unwanted metallic finds such as iron nails, silver paper, ring pulls, cola cans, and bottle tops. Often referred to in American literature as "trash".

L.C.D. - Liquid crystal display.

L.E.D. - Light emitting diode.

Loop - (See "Coil").

Masking - Where an unwanted find prevents a detector from responding to a wanted find (ie an iron nail buried in the ground close to a coin).

Meter Discrimination - Where a detector gives an audio signal for all finds but shows from meter response whether the target is wanted or junk. In its simplest form this is provided by a needle type meter, with the needle swinging left for junk or right for a good find. More complex versions give a number on an LCD meter to indicate the probable nature of a find, or even take the form of some type of graph on the meter display.

Mineralisation - Natural minerals in the ground or small fragments of iron on old habitation sites that cause loss of depth and erratic performance.

Motion - Detectors that can cancel ground effects but discriminate at the same time. To do this the search coil needs to be in movement. Most modern detectors work on the motion principle.

Negative Mineralisation - Loss of depth caused by ground contamination consisting of small particles of iron (on ancient habitation or industrial sites) or naturally occurring ferrous minerals. Normally encountered on inland sites.

Nicad - Nickel-cadmium rechargeable battery.

Non-Motion - Type of detector that registers a target even when the search head is held stationary over the top of it. Up until the early 1980s most detectors were of this type.

Notch - The facility on a detector to reject or accept items within a certain conductivity. For example such a detector could be set to accept only £1 coins or to reject all ring pulls of a certain type.

Null - A drop in threshold caused by the presence of iron or mineralised ground.

Overlap - Searching by moving forwards and sweeping the search head in increases of small amounts rather than by its full width.

Pinpoint - The ability of a detector to accurately indicate the precise position of a target buried in the ground.

Positive Mineralisation - Normally found on salt wet beaches, will cause a detector's set threshold to increase.

P.I. - Pulse Induction. A type of detector able to ignore mineralisation but usually, as a result of their sensitivity to iron, used for searching the salt wet sand of beaches.

Reject - The ability of a metal detector to ignore unwanted targets.

Search Head - (See "Coil").

Sensitivity - The depth at which a detector will register a metal object. Ideally, full sensitivity should be used for best depth, but in practice - as a result of mineralisation and other factors - more finds can often be made by turning the sensitivity down to match the conditions of the site.

Scattered Hoard - A collection of buried coins disturbed and scattered from their original place of deposition by ploughing or other causes.

Signal - Audio sound given by a detector when it locates a metal item.

Scuff Cover - Protective plastic plate, usually available as an accessory, which fits on the underside of the search coil to prevent scratching or other forms of damage. Usually individual to the make and model of detector concerned. Recommended as they can help maintain a detector's second hand resale value.

Silent Search - Most of the older-type detectors needed to be set to "threshold" (a faint but continual noise). Modern motion detectors run with a silent threshold apart from when a signal results from a target.

Stability - The ability of a detector's electronic circuitry to maintain threshold (silent or audible) without drifting out of tune. Instability can also result from outside electric interference (ie overhead power lines etc) or from "chatter" created by small pieces of iron or mineralised ground. Instability can be cured in many cases by reducing sensitivity or moving away from the source of the interference.

Stem - Tubular section of a metal detector used to connect the search coil to the control box.

Sweep - The way and speed at which a detector's search coil is moved from side to side while searching. The base of the search head should be kept parallel to the ground and as close to the ground surface as possible.

Target - Normally any buried metal item (although sometimes hot rocks or coke) that causes a metal detector to signal.

Target Masking - (see "Masking").

Threshold - The optimum point of tuning on a detector where a faint tone can just be heard. Most modern motion detectors do not need to be tuned in this way and can run on "silent threshold".

Tone Discrimination - The ability of a detector to indicate the likely nature of a target by a rise or fall in audio tone. Normally a drop in tone for junk, a rise in tone for wanted items.

Trash - (see "Junk").

Treasure Trove - Ancient law whereby precious metal objects, hidden with the intention of being retrieved, became the property of the Crown. Replaced in England, Wales and Northern Ireland by the Treasure Act 1996.

Treasure - In hobby terms, items covered under the Treasure Act 1996.

Tune - Adjust a detector's controls to facilitate best performance. Originally, to set a detector to threshold.

Variable Discrimination - A control on a detector that allows the level of rejection of junk items to be increased or decreased.

Visual I.D. - Where the meter of a detector indicates the probable nature of a target (see "Meter Discrimination").

Widescan - Double-D (or sometimes known as "2D") and other types of search coil capable of registering a target across their full diameter.

Britain's best-selling monthly magazine for
METAL DETECTING

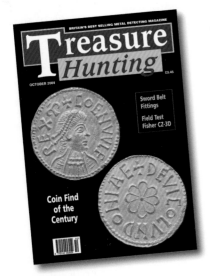

Every issue packed with useful information to help you get the most out of this fascinating hobby

- Site research
- Questions & Answers
- Field Tests on detectors
- "How To" articles
- Features on Coins & Artefacts found by detectorists
- Secondhand machine sales
- Detecting tales from around the UK with many illustrations of finds
- Ads from all the leading manufacturers & retailers
- Readers Letters ■ News & Views

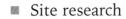

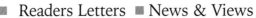

ON SALE AT

● Your newsagent ● Detector retailers ● Direct from the publishers - 01376 521900 (call for free sample copy)

www.treasurehunting.co.uk